AF849251

Self-Healing Sustainable Materials and Their Applications

Self-healing materials, inspired by biological systems, possess the unique ability to autonomously repair damage or respond to external stimuli, leading to enhanced durability and reliability of products. This book explores the interdisciplinary field of sustainable nanotechnology with a focus on self-healing materials, providing a comprehensive understanding of their synthesis, properties, and practical applications within the context of sustainability. It bridges the gap between scientific research and industrial application by presenting theoretical concepts alongside real-world case studies and implementations.

Features:

- Provides an in-depth exploration of the latest advances in sustainable nanotechnology, with a particular focus on self-healing materials
- Offers actionable insights into how self-healing materials can be integrated into various industries
- Adopts an interdisciplinary approach, covering the science, engineering, and sustainability aspects of self-healing nanomaterials
- Highlights how self-healing materials can contribute to sustainability by reducing waste, enhancing energy efficiency, and lowering long-term costs in manufacturing and maintenance
- Focuses on the environmental impact and sustainability considerations

This book is aimed at graduate students and researchers in materials science and nanomaterials.

Recent Advances in Materials for Nanotechnology (RAMAN)

Series Editor: Virat Khanna

"Recent Advances in Materials in Nanotechnology (RAMAN)" book series is a synthetic fusion of various domains of nanotechnology such as biological, chemical, physical, natural, agricultural, engineering, and medical sciences. This book series thoroughly covers the broad spectrum of nanotechnology applications and provides a vast array of information covering a wide range of nanotechnology fields to create an interface between research, academia, and industries. The main aim of this series is to compile a collection of monographs, handbooks, edited volumes, and original works that explore the complex nature of nanomaterials, nanodevices, and their manufacturing, characterization and applications.

Eco Friendly and Sustainable Coatings for Wear Prone Industrial Applications
Edited by Jimmy Mehta, Manikant Paswan and Pallav Gupta

Self-Healing Sustainable Materials and Their Applications
Edited by Ankit Tyagi, Akshay Dvivedi, Vijay Kumar Pal, and Vijay Chaudhary

For more information about this series, please visit: www.routledge.com/Recent-Advances-in-Materials-for-Nanotechnology/book-series/RAMAN

Self-Healing Sustainable Materials and Their Applications

Edited by

Ankit Tyagi, Akshay Dvivedi, Vijay Kumar Pal, and Vijay Chaudhary

CRC Press
Taylor & Francis Group
Boca Raton London New York

CRC Press is an imprint of the
Taylor & Francis Group, an **informa** business

First edition published 2027
by CRC Press
2385 NW Executive Center Drive, Suite 320, Boca Raton FL 33431

and by CRC Press
4 Park Square, Milton Park, Abingdon, Oxon, OX14 4RN

CRC Press is an imprint of Taylor & Francis Group, LLC

© 2027 selection and editorial matter, Ankit Tyagi, Akshay Dvivedi, Vijay Kumar Pal, and Vijay Chaudhary; individual chapters, the contributors

Reasonable efforts have been made to publish reliable data and information, but the author and publisher cannot assume responsibility for the validity of all materials or the consequences of their use. The authors and publishers have attempted to trace the copyright holders of all material reproduced in this publication and apologize to copyright holders if permission to publish in this form has not been obtained. If any copyright material has not been acknowledged please write and let us know so we may rectify in any future reprint.

Except as permitted under U.S. Copyright Law, no part of this book may be reprinted, reproduced, transmitted, or utilized in any form by any electronic, mechanical, or other means, now known or hereafter invented, including photocopying, microfilming, and recording, or in any information storage or retrieval system, without written permission from the publishers.

For permission to photocopy or use material electronically from this work, access www.copyright.com or contact the Copyright Clearance Center, Inc. (CCC), 222 Rosewood Drive, Danvers, MA 01923, 978-750-8400. For works that are not available on CCC please contact mpkbookspermissions@tandf.co.uk

For Product Safety Concerns and Information please contact our EU representative GPSR@taylorandfrancis.com. Taylor & Francis Verlag GmbH, Kaufingerstraße 24, 80331 München, Germany.

Trademark notice: Product or corporate names may be trademarks or registered trademarks and are used only for identification and explanation without intent to infringe.

ISBN: 978-1-041-06947-8 (hbk)
ISBN: 978-1-041-07245-4 (pbk)
ISBN: 978-1-003-63951-0 (ebk)

DOI: 10.1201/9781003639510

Typeset in Times
by SPi Technologies India Pvt Ltd (Straive)

Contents

About the Editors

Ankit Tyagi, PhD, Delhi Technological University, India. Ankit Tyagi is Associate Professor in the Department of Mechanical Engineering, Poornima College of Engineering, Jaipur, India. He got his PhD in manufacturing processes in advanced coating at Delhi Technological University, India. He visited various academic institutions and industries in several countries for academic pursuits. His area of research includes Surface Modification, Coating, Tribology, Material Processing, Composite Materials, Metal Matrix Nanocomposites, Nano-Coatings, Wear, Deformation, and Corrosion. He also has a good command of teaching subjects like welding technology, casting, wear mechanics, material science, non-destructive testing, metrology, and composite materials. He has published over 35 research papers in reputed journals and conference proceedings.

Akshay Dvivedi, PhD, IIT Roorkee, Uttarakhand, India. Akshay Dvivedi currently serves as Professor in the Mechanical and Industrial Engineering Department and holds the prestigious position of Dean, Sponsored Research and Industrial Consultancy (SRIC) at IIT Roorkee. His career is marked by significant contributions in both research and academic administration. He completed his PhD from IIT Roorkee in 2006, with an M.Tech. in Production & Industrial Systems Engineering (2004) from the same institute, and a Bachelor of Engineering from Jiwaji University, Gwalior (2000).

He has held several prominent positions throughout his career. As Associate Dean, Corporate Interaction (ADCI) at IIT Roorkee, he facilitated industry-academia collaborations. He currently leads the Automotive and Allied Research and Technology Innovation (AARTI) Foundation, IIT Roorkee's industry accelerator. He has also been the Coordinator of the Tinkering Lab, promoting innovation and hands-on learning among students, and serves as the Professor-in-Charge of the Advanced Manufacturing Processes Lab (AMP Lab) in the Mechanical and Industrial Engineering Department.

Under his leadership as ADCI and Dean SRIC, IIT Roorkee was adjudged as the "Most Innovative Research Institute" by the Confederation of Indian Industry (CII) for four consecutive years—2020, 2021, 2022, and 2023. This remarkable achievement highlights his efforts in promoting cutting-edge research and industry collaboration.

His research portfolio is extensive, with interests in unconventional machining processes, micro-machining, additive manufacturing, quality management, product design, and concurrent engineering. He has published over 150 papers in top-tier journals and conferences. He holds 7 granted patents, including innovations like a low-cost mechanical ventilator (also known as Pran Vayu) and a modular motorized accessory for manual wheelchairs.

He has supervised 19 PhD students and 40+ M.Tech. students, contributing to the development of the next generation of researchers and engineers. Prof. Dvivedi's efforts in research and education have solidified his standing as a key figure in both academic and industrial innovation at IIT Roorkee.

His commitment to education and research has garnered him several prestigious accolades, including the "Outstanding Teacher Award" from IIT Roorkee, which he has received twice. He was also awarded the "Corona Karmaveer Award" by the Chief Minister of Uttarakhand for his outstanding contributions to society during the COVID-19 pandemic. Furthermore, he was felicitated by the Hon'ble Governor of Uttarakhand for his exemplary leadership in the implementation of the TechSarathi Scheme at IIT Roorkee and for his invaluable support in promoting the growth of MSMEs.

Vijay Kumar Pal, PhD, IIT Kanpur, India. Vijay Kumar Pal currently serves as Assistant Professor in the Mechanical Engineering Department at IIT Jammu. His career is marked by significant contributions in both research and academic administration. He completed his PhD from IIT Kanpur in 2017. He visited various academic institutions and industries in several countries for academic pursuits. He has published more than 20 research papers in peer-reviewed international journals as well as in reputable international and national conferences. He has worked in four Sponsored Projects as a Project Investigator. His area of research is Stretchable electronics and sensors, Unconventional machining, Micro and nano manufacturing for hydrogen storage, Processing and Machining of Ultra high temperature ceramics, Tribology, etc.

Vijay Chaudhary, PhD, Netaji Subhas University of Technology, India. Vijay Chaudhary is currently working as Assistant Professor in the Department of Mechanical Engineering, Amity University Uttar Pradesh, Noida. He has completed his PhD in Mechanical Engineering from the Netaji Subhas University of Technology, University of Delhi. He has published more than 100 research papers in peer-reviewed international journals as well as in reputable international and national conferences. He has edited several books with international publishers. Dr. Chaudhary has over 8 years of teaching and research experience. He has a good command of teaching subjects like Material Science; Engineering Materials; Manufacturing Processes; Measurement and Metrology & Composite Materials. Currently, he is working in the field of biofiber-based polymer composites. His research area of interest lies in the area of Processing and characterization of Polymer composites, Tribological analysis of biofiber-based polymer composites, water absorption of biofiber-based polymer composites, and surface modification techniques related to polymer composite materials. Based on his research profile, Dr. Chaudhary presently holds h-index of 15 and i10-index of 20 as per Google Scholar.

Contributors

Trapty Agarwal
Maharishi School of Engineering and Technology
Maharishi University of Information Technology
Uttar Pradesh, India

Chanchal Ahlawat
SCSET, Bennett University
Greater Noida, Uttar Pradesh, India

Mehran Khan Alam
School of Material Science and Engineering
Shandong University
250061 Jinan, China

Gynanedra Prasad Bagri
Department of Mechanical Engineering
Faculty of Engineering and Technology
SRM Institute of Science and Technology Delhi NCR Campus
Modinagar, India

Masum Bellah
Department of Mechanical Engineering
University of Wisconsin – Milwaukee
Milwaukee, USA

T. Bernatin
Department of Electronics and Communication Engineering
Sathyabama Institute of Science and Technology
Chennai, India

Anant Bhardwaj
Department of Mechanical Engineering
Poornima College of Engineering
Jaipur, Rajasthan, India 302022

Vijay Chaudhary
Department of Mechanical Engineering
A.S.E.T, Amity University
Uttar Pradesh, Noida, India

Rina Kamlesh Chokshi
Department of Civil Engineering
Faculty of Engineering and Technology
Parul Institute of Engineering and Technology
Parul University, Vadodara, Gujarat, India

Kabir Dhingra
Department of Mechanical Engineering
A.S.E.T, Amity University
Uttar Pradesh, Noida, India

Weiguang Fan
School of Mechanical Engineering
Shandong University
Jinan 250061, PR China

Liu Feifei
School of Mechanical Engineering
Shandong University
Jinan 250061, PR China

Jay Gandhi
Department of Computer Science and Engineering
Faculty of Engineering and Technology, Parul Institute of Engineering and Technology
Parul University
Vadodara, Gujarat, India

V. Haripriya
Department of CSIT, Jain (Deemed-to-be University)
Bangalore, Karnataka, India

Aashay Ingle
School of Mechanical Engineering
Shandong University
Jinan 250061, PR China

Harshit Joshi
Department of Mechanical Engineering
A.S.E.T, Amity University
Uttar Pradesh, Noida, India

Harshu Kumar
Department of Mechanical Engineering,
A.S.E.T, Amity University
Uttar Pradesh, Noida, India

Awakash Mishra
Maharishi School of Engineering and Technology
Maharishi University of Information Technology
Uttar Pradesh, India

Ghulam Mustafa
School of Mechanical Engineering
Shandong University

Praveen Pachauri
Principal, Government Polytechnic Siwan
Department of Science, Technology & Technical Education
Patna, Bihar, India

A. Prabhu
Department of CSIT, Jain (Deemed-to-be University)
Bangalore, Karnataka, India

T. Ravi
Department of Electronics and Communication Engineering
Sathyabama Institute of Science and Technology
Chennai, India

Pradeep Rohatgi
Department of Mechanical Engineering
SR University
Warangal, India

Rohit Sahu
Department of Mechanical Engineering
G.L Bajaj Institute of Technology and Management
Greater Noida, U.P. India

Savita
Maharishi School of Engineering and Technology
Maharishi University of Information Technology
Uttar Pradesh, India

Sumit Sharma
Department of Mechanical Engineering
A.S.E.T, Amity University
Uttar Pradesh, Noida, India

Deepak Singh
Department of Mechanical Engineering
I.E.T., Lucknow, India

Sachin Sirohi
Department of Mechanical Engineering
Faculty of Engineering and Technology
SRM Institute of Science and Technology
Delhi NCR Campus, Modinagar, India

V. Sivachidambaranathan
Department of Electrical and Electronics Engineering
Sathyabama Institute of Science and Technology
Chennai, India

Vaibhav Srivastava
Department of Materials Science and Engineering
University of Wisconsin – Milwaukee
Milwaukee, USA

Ankit Tyagi
Department of Mechanical Engineering
Poornima College of Engineering
Jaipur, Rajasthan, India 302022

Falguni Tlajiya
Department of Electrical Engineering, Faculty of Engineering and Technology, Parul Institute of Engineering and Technology
Parul University, Vadodara, Gujarat, India

Chenglong Yang
School of Mechanical Engineering
Shandong University
Jinan 250061, PR China

Preface

This book examines the rapidly advancing domain of sustainable nanotechnology, emphasizing the potential application of self-healing materials in mitigating environmental challenges. Self-healing materials, inspired by nature's regenerative characteristics, are increasingly favored for their capacity to autonomously repair damage, extend product longevity, and mitigate waste and resource depletion. The book offers a comprehensive analysis of the creation, application, and environmental impacts of these materials across several industries by integrating sustainability principles with nanotechnology. This discussion will encompass the latest advancements in research, practical industrial uses, and the potential of these materials to enhance environmental stewardship, diminish carbon emissions, and strengthen the circular economy. The book aims to serve as a valuable resource for scholars, professionals, researchers, and politicians committed to sustainable innovation and materials science, featuring comprehensive case studies, empirical research, and progressive perspectives.

Scientists, engineers, and policymakers have been motivated to reevaluate the design and application of materials due to the growing need for environmentally sustainable practices. Nanotechnology has significantly enabled the development of advanced materials with exceptional functionality. Self-healing materials represent a transformative innovation in this domain, since they have the potential to enhance durability, reduce waste, and improve energy efficiency. This book examines the interdisciplinary integration of self-healing systems with nanotechnology. These materials emulate biological healing processes, yielding more dependable and durable products by being engineered to autonomously or in response to external stimuli repair structural or functional damage. Their performance improves and their application broadens in sectors like as renewable energy, electronics, healthcare, transportation, and infrastructure when nanoscale technologies are integrated into their design. The primary objective is to demonstrate a thorough understanding of the synthesis methods, key properties, and practical uses of self-healing materials within a sustainability context. The book connects laboratory research with large-scale applications by integrating theoretical foundations with practical examples and advanced industrial methods. It offers an extensive perspective on how self-healing nanomaterials might advance the global objective of sustainable development by examining previous achievements and forecasting future trajectories.

1 Introduction to Sustainable Nanotechnology

Trapty Agarwal, A. Prabhu, V. Sivachidambaranathan, Jay Gandhi, Rohit Sahu, Vijay Chaudhary, Liu Feifei, and Ghulam Mustafa

1.1 OVERVIEW OF NANOTECHNOLOGY

The establishment of pollution-free technology and renewable energy sources is crucial for sustained human development. Despite ongoing challenges such as waste proliferation, carbon emissions, and the depletion of nonrenewable resources, innovative technological advancements are continually being developed. Civilization is shifting toward a greener future to tackle the negative impact of technology on individual well-being and the atmosphere. Nanotechnology has been utilized in various areas in recent decades (Del Prado-Audelo et al., 2021).

The term **nano** comes from the Greek term, denoting **extremely tiny**. Nanoscience is the study of structures with atomic elements. It brings material science, chemistry, and biology together. Nanoscience is described as:

- The study of atoms and substances that contain at least one element in the nanoscale (10^{-9} m) range.
- The understanding of how to design and manage structures using Nanomaterials (NMs).

Nanoscience is an integrated discipline that requires association among biologists, chemists, and physicists to address scientific concerns (Desmond et al., 2021).

- Nanotechnology is the application of nanoscience. It involves designing, synthesizing, characterizing, and using matters, appliances, and systems at the nanoscale.
- The development and utilization of nanoscale materials systems.

Nanotechnology is concerned with the production of substances with high accuracy. The distinction between nanoscience and nanotechnology (NS&NT) is similar to the

DOI: 10.1201/9781003639510-1

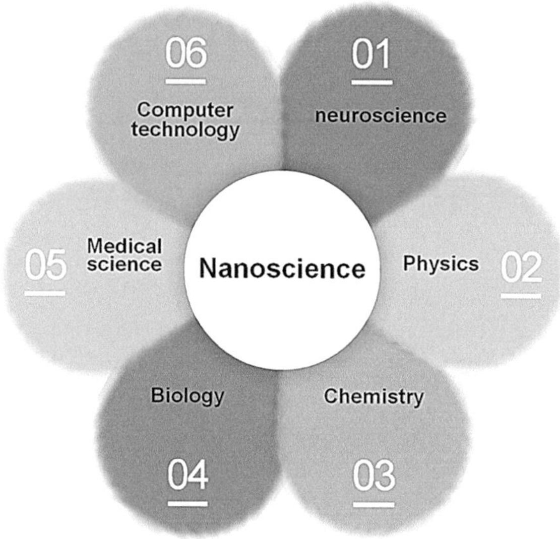

FIGURE 1.1 Connection between nanoscience and other fields.

distinction between science and application, although both require extreme accuracy. These materials are constructed in layered, massive, and innovative structures through various fabrication methods (Kuperkar et al., 2024). These materials are recognized as NMs. Figure 1.1 shows the relationship between nanoscience and other fields.

NMs have been created to be smart and versatile, resulting in fresh and greener inventions with important well-being and environmental advantages over the past several years. The NM offers several applications, including medicine and pharmaceuticals, battery nanostructured electrodes, electronic data storage devices, and antibiotic compounds in the cosmetics, food, and textile sectors. Nanotechnology applications aim to reduce fuel consumption, energy use, waste generation, and freshwater pollution; promote recycling; eliminate plastic bag usage; and support renewable energy sources and environmental responsibility (Zhang et al., 2021). The researchers describe different morphologies, including fibers, rods, wires, tubes, spinel, petal-shaped, spherical, floral form, trigonal, cubic, hexagonal, tetrahedral, decahedral, octahedral, and polyhedral. NMs exhibit distinct characteristics, including electrical, magnetic, mechanical, catalytic, and optical. Nano-substances' small size, large surface area, and unique physicochemical features make them very sensitive, potentially harming both biotic and abiotic ecosystem components. As a consequence, increased stress is placed on public health and environmental safety (Harris et al., 2023).

In 2004, the Royal Society and Royal Academy of Engineering issued the first description of nanotoxicity, emphasizing the importance of studying and inspecting the toxicity of NMs at various levels, including biological, molecular, and genetic, to ensure their safety before use. Although numerous studies have been conducted on physical contamination, oxidative damage, inflammation, cell death, and DNA damage of nano-substances, there is still insufficient data and a need for more inquiry (Kiran et al., 2022).

1.2 HISTORY OF NANOTECHNOLOGY

The historical development of nanotechnology remains somewhat unclear, although Richard Feynman is widely regarded as one of its early conceptual pioneers. Before Feynman, nanoparticles were already employed for several uses, including medicine. John Utynam copyrighted gold (Au) NM-centered glass in 1449 (Jagtiani, 2022). In the 16th decade, Swiss physicians often recognized as Paracelsus, employed Au nanoparticles to cure numerous diseases. Although the area of NS&NT is not innovative, its methodologies have evolved with period. Richard Zsigmondy, who got the Nobel Prize in Chemistry in 1925, was the first to develop the notion of nanometers. They utilized a microscope to determine the size of gold colloidal atoms. Feynman did not coin the term **nanotechnology** but did build on it in a 1959 talk at a global symposium. In 1974, Norio Taniguchi coined the word **nanotechnology** to characterize the practice of nano-sized substances and their precision at the nanometer (Kumar et al., 2021). The discovery of groups in particle knowledge and the development of the Scanning Tunneling Microscope fueled advancements in nanotechnology.

This string of breakthroughs continued, and fullerenes and Carbon Nanotubes (CNT) were discovered very recently. These breakthroughs paved the way for new research pathways and led investigators and researchers to fresh fields ripe for the application of nanotechnology. In the US, two projects were launched in 1991 and 2001 to raise understanding about the probable uses of nanotechnology. The global center for knowledge appraisal assessed nanotechnology development in several nations, in addition to previously listed projects. In 1999, the Interagency Working Group on Nanoscience (IWGN) hosted a meeting to anticipate nanotechnology investigation during the following ten years (Modi et al., 2021). The Presidential Council on Science formally endorsed the IWGN's results and recommendations in 2000.

1.3 FOUNDATIONAL PRINCIPLES OF NANOTECHNOLOGY

NMs possess unique features that distinguish them from larger-scale materials. NM characteristics differ greatly from bulk substances due to two factors: surface-to-volume ratio (SVR) and quantum effect (QE). The following are the primary factors that impact property changes.

1.3.1 Quantum Effect

The development of nanostructured materials is the most important development of this period. The QE is a key characteristic distinguishing nanoscience from former sciences. The QE pertains to quantum mechanics and is crucial for understanding atomic structure. The Bohr radius is a key component in observing the quantum size effect (Sindhwani & Chan, 2021). When the energy level splits below the Bohr radius, the visual band gap expands rapidly, resulting in NMs with an enormous simulation radius exhibiting QE. As particle size decreases, energy levels become more dispersed. A drop in size leads to a fast rise in the surface area, resulting in a QE that alters material characteristics.

Quantum confinement refers to the spatial detention of electron-outlet duos in single or more dimensions of a substance. Semiconductors exhibit QE due to their electronic structure, which has a band gap that metals lack. Quantum confinement occurs when the material size is decreased to 2 nm in all dimensions. Moving atoms closer together creates a continuous band of states. As the number of atoms decreases, delocalized energy levels become spatially limited. When electrons' de Broglie wavelength and particle diameter match, electron-hole pairs are confined in the conduction band. Nanostructures are predicted to exhibit stronger QE when particle diameters decrease below Bohr's radius. It is vital to remember that several elements, including barrier layers and band offset, contribute to the variance in QE. In quantum confinement, the probabilistic nature of particles enables them to tunnel through potential barriers without the need for additional energy.

The characteristics of materials vary dramatically when particles go from the macro-scale to the nanoscale. Power conditions appear constant initially, but become distinct as restricting dimensions shrink. As power positions shift from constant to separate, NMs and their structures become more form and size-dependent. As the size and structure of an NM decrease, optical lighting shifts from red to blue. QE discusses the events that occur when excitons are confined in elements exceeding the Bohr radius. Quantum computing and nanostructures need accurate computations to learn from several quantum procedure literatures. Bohr's radius is an $a_B = \frac{h}{m_c C\alpha}$. h is the condensed Planck's factor, C is the velocity of the beam, α is the well formation variable, and m_c is the electron's rest mass. The QE produces quantum dots for practical applications. The QE hypothesis is based on the particle-like behavior of the simulation due to its reduced element.

Material dispersion varies with dimension, revealing data regarding the bulk conditions (stable in 2D), the inverse cube root dependence of energy, and comparable sharing in valence and transmission bands. Nano-sized hetero connection devices featuring QE are suitable for many applications in electronics, optoelectronics, and physics. Wurtzite nitride quantum walls hetero-structures are effective for light-emitting- and laser diode applications, particularly in quantum dots. One-dimensional quantum well-wire structures have received significant interest for their competent physical characteristics and applications in optoelectronic nanodevice systems. These structures, such as wurtzite GaN and ZnO NM quantum structures, exhibit unique properties at the nanoscale. Quantum-confined nanostructures thrive in practical applications, demonstrating their efficiency and versatility.

1.3.2 Surface-to-Volume Ratio (SVR)

SVR in nanoparticles significantly influences their properties. To begin with, NMs have a significantly higher surface area than an equal amount of material. For example, take a sphere with a radius r. NMs and their structures exhibit unique features due to their enormous surface area, which is influenced by the material's size, shape, and atoms. To illustrate this, let us use cubic atoms as an example to examine how dimension influences the outside quarters in nanostructures. To demonstrate, a 1 μm cube is cut into smaller cubes with edge lengths of 100 or 10 nm. Converting huge materials to smaller ones does not result in a loss of mass. Figure 1.2 represents the calculation of surface area by dividing the cube.

Nanotechnology relies heavily on the inverse relationship between surface area and particle size, resulting in a decrease in facade area as atom size enlarges (Abdalla et al., 2022). Use the mass equilibrium Equation (1.1) to calculate the number of smaller cubes and assume no material loss as the volume changes from large to small.

$$N_c = \left[\frac{L}{l}\right]^3 \tag{1.1}$$

N_c denotes the number of small cubes created by splitting the relative cube into pieces with varying edge lengths. The smaller cubes' volume and edge length are denoted by L and l, respectively. After generalization, Equation (1.2) yields the number of cubes.

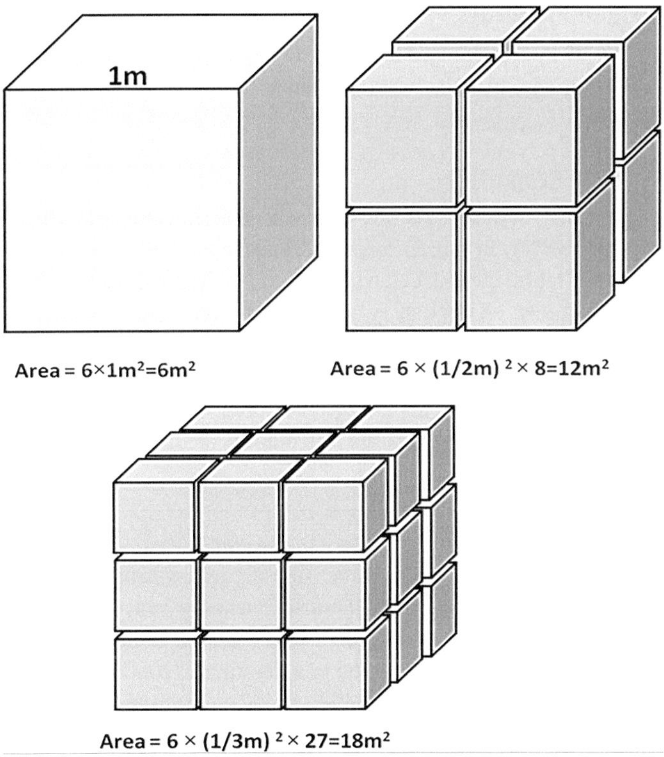

FIGURE 1.2 Computation of SVR.

$$N_c . \frac{\alpha}{A} = N_c \left[\frac{L}{l}\right]^2 \quad (1.2)$$

Equations 1.1 and 1.2 show that when the mass of a cubical atom decreases from volume to nano, the surface area increases proportionally. In the equation above, α denotes the surface region of a smaller cube. Subdividing the relative cube into smaller cubes with edge durations ranging from 0.01 to 1 micron improves the relative surface region. As particle size decreases from micro to nanoscale, surface area and particle density increase per unit mass. Smaller divisions of a cube increase surface area, demonstrating the potential of nanotechnology.

This indicates that when the cube decreases, the SVR increases, and vice versa. It is also being concluded that dividing a given volume into minor parts increases the facade area. As the atomic dimension decreases, a greater proportion of the atoms reside on the surface rather than inside. As a result, nanoparticles have significantly more surface area per unit than larger atoms. It makes NMs more chemically reactive. Since growth and catalytic reactions happen at surfaces, a known mass of NM is far more reactive than an equivalent mass of bulk material. It has been discovered that materials considered inert in bulk form often become reactive at the nanoscale, thereby enhancing their properties.

1.3.2.1 Unique Properties

NMs exhibit unique physical, mechanical, thermal, chemical, electronic, and magnetic characteristics compared to bulk materials. Here are some key features of nanoscale materials.

1.3.2.2 Physical Properties

NMs have the same crystal structure as bulk materials due to differing lattice characteristics. However, in nanoform, the crystal structure is icosahedral. Lithium has a face-centered tetrahedral structure and is considered nanoscale when the size is less than 6.5 nm in size (Asamoah et al., 2024). NMs have less interatomic space than bulk materials due to stronger surface cohesive forces, along with enhanced van der Waals forces and metallic bonding at the nanoscale. As the interatomic gap in Al drops from 2.86 Å to 2.81 Å, the required force reduces from 3.39 to 2.77 eV (Reents et al., 2025). Nanostructured substances exhibit exclusive crystalline properties.

1.3.2.3 Thermal Properties

Thermal characteristics change as the dimension decreases near the nanoscale. NMs' melting properties decline with greater surface energy and inter-atomic spacing changes. The melting point of a material refers to the temperature increase that occurs when it transitions from an ordered to a disordered condition. Metals less than 30 nm exhibit a gradual reduction in melting point. Solid atoms oscillate about their mean locations, and the amplitude of these vibrations rises with temperature (Liu et al., 2021). When vibrations surpass the bond length, the melting point propagates from the surface to the interior. Atoms within the crystal lattice have less freedom to

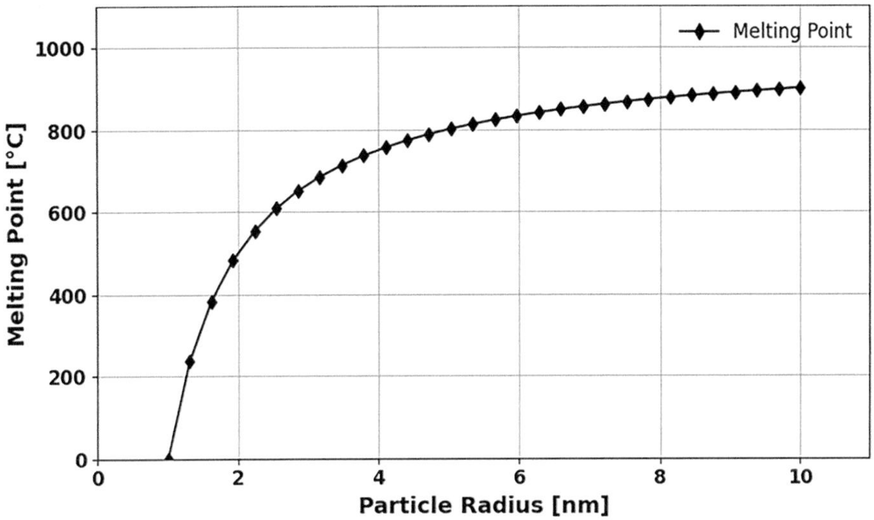

FIGURE 1.3 Atom dimension compared to temperature gradient for gold nanoparticles.

vibrate compared to those at the surface and grain borders. As grain size decreases, more atoms are concentrated near the surface and borders. Figure 1.3 illustrates that the melting point of gold nanoparticles falls with size.

1.3.2.4 Mechanical Properties

NMs have many flaws during manufacture, affecting their mechanical characteristics. Nanostructures, such as CNT exhibit unique properties such as superior mechanical force, elastic edges, and elasticity compared to their bulk structure (Ciobanu & Aradoaei, 2024). Nano-phase stone wares exhibit flexibility at extreme temperatures, while granular porcelain is fragile.

1.3.2.5 Magnetic Properties

NMs with high SVR reveal separate magnetic characteristics compared to bulk substances. A ferromagnetic material has numerous magnetic domains, but at the nanoscale, only one domain shows superparamagnetic events. Nanoparticles have a significantly lower magnetic moment compared to their bulk size.

1.3.2.6 Optical Properties

The transition barrier between the valence and conduction bands determines optical characteristics in nano-sized materials. Low-scaled semiconductor and metal materials exhibit significant changes in optical properties, including color. Au nanoparticle colloidal substances have a profound crimson hue that transitions to yellow as particle size increases, resulting in a façade Plasmon character in nanoscale substances (Ebrahim et al., 2024). Nano-sized semiconductors exhibit size-dependent features that vary with light intensity and frequency. They also improve nonlinear optical properties and increase emission energy or wavelength. The size factor affects several optical characteristics, including photoconductivity, photocatalysis, photoemission, and electroluminescence.

1.3.2.7 Electronic Properties

Electronic characteristics are connected to the wave-like behavior of electrons in small-scale materials. The material size approaches the de Broglie wavelength, and the energy levels become more distinct. Below the crucial length scale, the conduction material widens its bandgap and becomes an insulator. Nanoparticles containing semiconducting elements such as Ge, Si, and Cd are not semiconductors (Arabczyk et al., 2024). Preparing nanoparticles of TiO_2 by precipitation yields straight band-gap semiconductor activity, while microcrystalline TiO_2 provides an indirect band-gap substance.

1.4 INNOVATION AND APPLICATIONS

NMs are establishing themselves as revolutionary catalysts in a variety of sectors due to their distinctive nanoscale characteristics. Their novel uses range from power generation to ecological cleanup, allowing for improvements previously impossible with ordinary substances. Unique innovations and applications are given below:

- Energy efficiency

  Energy generation and storage are crucial for meeting global energy demands while promoting sustainability and minimizing environmental consequences. Energy production includes creating electricity from diverse sources, while energy storage devices store excess energy for future use. NMs offer a vital function in improving the efficiency, capacity, and sustainability of energy systems, as renewable energy and energy efficiency become increasingly important.
- Solar energy

  NMs, such as quantum dots, nanowires, and graphene, improve light absorption and electron transport and reduce energy losses in photovoltaic cells (solar panels). Perovskite solar cells are gaining popularity due to their high efficiency and inexpensive manufacturing costs. Nanostructured solar thermal collectors use NMs (Gottardo et al., 2021). CNT and nanofluids are utilized in solar thermal systems to enhance heat absorption and transmission, making solar energy generation more efficient.
- Wind energy

  Nanostructured materials, such as carbon fiber composites and graphene, are being employed in wind turbine blades to improve strength, durability, and lightweight, leading to increased energy efficiency and longer turbine lifespan (Aydin et al., 2023).
- Waste reduction

  Waste minimization is a primary goal of sustainable nanotechnology. Catalytic reactions help to cut thermal energy consumption and waste creation dramatically. Catalysts are used in small amounts to perform a reaction several times. They outperform stoichiometric reagents by reacting selectively or not at all under typical circumstances. It aims to reduce or eliminate waste and related side products, as prevention is better than a cure. This requires a change from concentrating on the element produced to prioritizing waste reduction (Kuhn et al., 2022). The nuclear economy idea emphasizes developing synthetic techniques to maximize the production of end goods by all process factors. For instance, when the reduction in ketone diminution to minor alcohol utilizing Sodium Borohydride (SBH) and H^+ is evaluated, it is found that SBH has an 81% nuclear wealth, whereas the H^+ direct has 100% atom economy, which denotes the whole invention is used and no waste is produced.

- Pollution remediation

  Nanoremediation uses nanoparticles to clean and eliminate toxins from the ecology, including soil, water, and air. NMs' unique features, including high reactivity, huge surface area, and improved adsorption, enable efficient degradation, neutralization, and removal of contaminants. Nanoremediation is effective in removing environmental contamination from heavy metals, organic pollutants, and persistent toxins. Nanotechnology plays a crucial role in remediating polluted water, both on-site and off-site. NMs' cleanup is effective as it reaches areas where bigger particles cannot, and offered a higher surface-area-to-volume ratio for improved contaminant adsorption. Furthermore, effectively coated NMs prevent interactions with undesirable substances in the media (Kumar, 2023). Chlorinated solvents from hazardous waste sites have been effectively removed from groundwater through treatment. The use of zerovalent iron NMs for quicker dechlorination of perchloroethylene and trichloroethylene resulted in the production of carbon dioxide. Nanoscale iron particles are effectively used to immobilize lead, arsenic, and chromium from contaminated waste sites.

- Water purification

  NMs have transformed water treatment with their large outside region, porosity, and reactivity at the nanoscale. NMs purify water by removing heavy metals and organic pollutants, disinfecting, and sterilizing. Using NMs in water filtration has several benefits, including enhanced efficiency, cost savings, and safety. Nanofiltration membranes are semipermeable membranes used for water treatment and purification. These filters, with particle diameters ranging from 1–10 nm, effectively remove pollutants such as dissolved salts, bacteria, organic pollutants, and heavy metals from water. The nano-membrane has a molecular weight cut-off of 200–1000 Da, high water permeability for effective treatment, and excellent impurity and contaminant rejection efficiency (Liu et al., 2024). Nanoparticles, including CNT, graphene oxide, and silver nanoparticles filter, adsorb, and eliminate pollutants from water, such as heavy metals, organic pollutants, and pathogens.

- Air pollution control

  Air pollution control involves reducing dangerous pollutants in the air to safeguard human health, the environment, and ecosystems. Common pollutants include PM, NOx, SO_2, CO, VOCs, and greenhouse gases (CO_2 and

CH_4). Effective air pollution control encompasses both preventative and treatment measures to minimize emissions from industrial operations, transportation, power generation, and other sources.

- Air pollution control technologies are divided into two categories: controlling particulate matter and controlling gaseous pollutants (Roy et al., 2021).
- Source Control: Adopting cleaner technology directly at the pollution source significantly reduces emissions. Examples include employing low-emission fuels in vehicles and sophisticated combustion processes in industrial furnaces.

Regulations often need companies to implement pollution control systems or adopt cleaner methods of operation to meet emission standards.

- Sustainable manufacturing

  Nanotechnology allows for the precise manufacture of nanoscale components, leading to improved manufacturing processes and product quality. Nanoscale components enable device and system downsizing, resulting in smaller, more efficient products. Researchers are now working to produce nano-steel. Arcelor Mittal is using NM-based steel to create softer, thinner columns and panels at the lowest price. Steel columns and panels have approximately the same mass as aluminum. CNT-implanted substances have the potential to be used in numerous construction tools. Buckyballs are tiny nanoparticles used in badminton rackets to enhance striking energy and consistency. Using lighter, thinner dye coats on airplanes is improving flying efficiency (Li et al., 2023).
- Green Nanotechnology

  Green nanotechnology, a key area of environmental science, promotes the development of clean and sustainable technology. This technology combines green chemistry and engineering ideas to protect the environmental impact and reduce waste. It has the potential to significantly affect several industries, including medicine, agriculture, the environment, nanotechnology, and electronics. Advancement in quantum mechanics and nanotechnology-based devices could profoundly altar society. As electricity becomes the primary energy source, leading to a fast shift in consumption patterns (Popescu & Ungureanu, 2023). Energy storage technologies are driving transformation in the electronics industry, particularly in stationary applications, where advanced bulk energy storage systems, such as those used in control panels and electric vehicles, are making an impact. Nanotechnology has successfully improved energy storage in batteries and super-capacitors. Innovations such as ceramic separators and enhanced electrode materials have the potential to significantly improve the power and safety of lithium-ion batteries, crucial for EVs and grid-level storage, both vital for renewable energy integration. The performance of renewable lithium batteries is largely determined by the physicochemical properties of the electrode materials. New materials based on the Prussian blue analogy are being studied for grid-size functions. For example, a manganese hexa-cyano manganate anode paired with a copper hexa-cyanoferrate cathode has

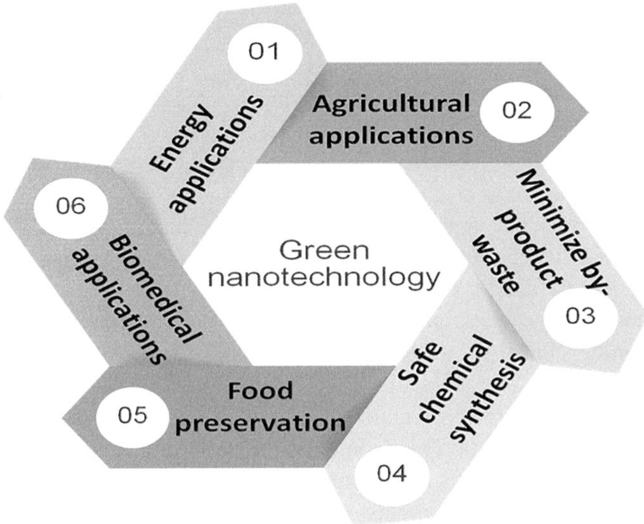

FIGURE 1.4 Applications of green nanotechnology.

resulted in a secure, fast, and economical aqueous battery with no observable capacity loss after 1000 deep freeze cycles. Figure 1.4 illustrates the applications of green nanotechnology (Kumar et al., 2024).

1.5 IMPACTS OF NANOTECHNOLOGY

While nanotechnology has many benefits, it also presents certain drawbacks and obstacles.

Several drawbacks and issues with nanotechnology are listed below:

- Nanoparticles pose ecological problems due to their unique behavior and toxicity compared to larger particles. This has risks in both aquatic and terrestrial ecosystems.
- Nanoparticles pose health risks when inhaled, ingested, or absorbed through the skin, raising concerns about occupational safety and consumer exposure.
- Nanotechnology raises ethical and regulatory concerns, including human augmentation, surveillance, and misuse in areas like weapons development.
- The long-term effects of nanoparticles on human health and the environment are still unknown due to their nascent stage of nanotechnology. Research is addressing these knowledge gaps, but more data is needed to understand the full extent of the risks.
- The high development and implementation costs of nanotechnology limit its use, particularly in sectors or regions with fewer resources. Ensuring reasonable contact to nanotechnological innovations remains a significant challenge (Winiarska et al., 2024).

- NM toxicity varies depending on their morphology and surface chemistry. This heterogeneity composes it difficult to fully measure their security and regulatory needs.
- Nanotechnology presents principled concerns regarding potential misuse, especially in surveillance, cyber intrusions, and human enhancement. Striking a balance between technological advancement and security is a significant challenge.
- The production, use, and disposal of products containing NMs unintentionally release nanoparticles into the atmosphere, possibly influencing ecologies and individual well-being.
- Nanotechnology raises ethical concerns that must be carefully addressed. Developing ethical principles and frameworks for research and development is a continuous problem.
- The benefits and risks of nanotechnology are not yet fully understood. Valuable communication and communal learning are essential for making knowledgeable decisions and using nanotechnology responsibly.
- Nanotechnology developments often rely on inventive research and improvement. Intellectual property disputes hinder evolution and access to essential technologies, limiting the potential benefits (Okem et al., 2024).

Although nanotechnology presents certain challenges, in progress research, guidelines, and ethical frameworks are developing to tackle these issues, maximize benefits, and minimize risks. The responsible advancement of nanotechnology requires careful ethical deliberations and rigorous estimation of possible implications.

1.6 ROLE OF GREEN CHEMISTRY

Green architecture and cost-effective eco-manufacturing are the key drivers for substantial technological renewal and longevity. The goal is to enhance energy efficiency, minimize pollution and greenhouse gas emissions, and reduce the use of fossil fuels such as feedstock. Green chemistry attempts to enhance sustainable production and reliable manufacturing while decreasing waste creation. However, numerous restraints change in the adoption of green production methods, and the potential of employing nanoscience is accelerating the technical enterprises in green nanotechnology and large-scale efficient synthesis (Kong & Li, 2021). Below are some key recommendations for the near future:

- The initial stage in green chemistry is to investigate the applications and inferences of green production and techniques. This holistic method of technology development encourages multidisciplinary research and effective communication. Engaging all stakeholders is crucial for the success of any new technology since public expectation and acceptance are key factors.
- Improving the large-scale manufacturing of NMs is crucial for realistic functions and trade development. This includes developing new synthesis

techniques and advanced manufacturing equipment to support green innovation.

- Guidelines for producing synthetic NMs should encourage enterprises to use green nanotechnology-based production. Additionally, testing the efficiency of combining green and synthetic NMs is recommended to optimize performance and sustainability.
- The uncertainties surrounding green nanoparticles require serious examination. This is only possible if academic institutions and focused research are involved and supported (Nair et al., 2022). Toxicologists' assistance helps to bridge the gap between green invention and its potential health and environmental consequences.

1.7 CONCLUSION

NMs have transformed industries such as water purification, air purification, soil remediation, and medicinal applications due to their unique properties, including high surface area, adjustable chemical reactivity, and multifunctionality. While their integration into various sectors presents intriguing answers to global concerns, it is important to examine the associated hazards. NMs like CNT, graphene, and metal oxides efficiently eliminate impurities from water. Water treatment technologies have improved due to their capacity to degrade or adsorb toxic compounds at the molecular level, resulting in safer and more environmentally friendly water supplies. Nano-catalysts and photo-catalysts effectively remove pollutants, including nitrogen, sulfur, and volatile organic compounds, from the air. These components help neutralize pollutants and significantly improve air quality in urban and industrial areas. Nanotechnology has improved soil remediation by immobilizing and converting hazardous chemicals such as pesticides, heavy metals, and hydrocarbons. Deploying them on polluted soil restores ecological balance and mitigates pollution's long-term effects on agriculture and biodiversity.

NMs have revolutionized biomedical applications, including medication administration, imaging, diagnostics, and tissue engineering. Nano-carriers enhance medication distribution, reduce side effects, and enhance treatment efficiency. Nanoparticles have revolutionized drug delivery, imaging, diagnosis, and tissue engineering. Nano-carriers enhance targeted medication delivery, reduce side effects, and improve treatment efficacy. Nanoparticles facilitate early detection through advanced imaging techniques, while biosensors and nanotechnology are transforming modern healthcare.

In conclusion, NMs have emerged as revolutionary instruments in the water, air, soil, and medicinal sectors, providing novel and highly efficient answers to global concerns. While their unique features offer enormous advantages, careful consideration of environmental and health consequences remains critical. Nanotechnology is accomplishing long-term development by encouraging multidisciplinary cooperation, supporting green synthesis, and conducting rigorous risk assessments. This will drive future advancements while protecting human and environmental health.

Chemical symbols	Abbreviations
PM,	Particulate matter
NOx,	Nitrogen oxides
SO_2,	Sulphur dioxide
CO,	Carbon monoxide
VOCs,	Volatile organic compounds
CO_2	Carbon dioxide
CH_4	Methane
H^+	Hydrogen
TiO_2	Titanium dioxide
Ge	Germanium
Si	Silicon
Cd	Cadmium
Al	Aluminum

REFERENCES

Abdalla, J. A., Thomas, B. S., Hawileh, R. A., Yang, J., Jindal, B. B., & Ariyachandra, E. (2022). Influence of nano-TiO_2, nano-Fe2O3, nano clay, and nano-CaCO3 on the properties of cement/geopolymer concrete. *Cleaner Materials*, *4*, 100061. https://doi.org/10.1016/j.clema.2022.100061

Arabczyk, W., Pelka, R., Brzoza-Kos, A., Kocemba, I., Rokicka-Konieczna, P., Skulmowska-Polok, K., . . . & Lendzion-Bieluń, Z. (2024). studies of phase transformation kinetics in the system of nanocrystalline iron/ammonia/hydrogen at the temperature of 350° C using magnetic permeability in situ measurement. *Applied Sciences*, *14*(18), 8452. https://doi.org/10.3390/app14188452

Asamoah, G. A., Korsah, M., Jeyasundar, P. G. S. A., Ahmed, M., Lau, S. Y. and Danquah, M. K., 2024. Nanotechnology-based lithium-ion battery energy storage systems. *Sustainability*, *16*(21), 9231. https://doi.org/10.3390/su16219231

Aydin, M., Koc, P., & Sahpaz, K. I. (2023). Investigating the EKC hypothesis with nanotechnology, renewable energy consumption, economic growth and ecological footprint in G7 countries: Panel data analyses with structural breaks. *Energy Sources, Part B: Economics, Planning, and Policy*, *18*(1), 2163724. https://doi.org/10.1080/15567249.2022.2163724

Ciobanu, R. C., & Aradoaei, M. (2024). Techniques and instruments for assessing and reducing risk of exposure to nanomaterials in construction, focusing on fire-resistant insulation panels containing nanoclay. *Nanomaterials*, *14*(18), 1470. https://doi.org/10.3390/nano14181470

Del Prado-Audelo, M. L., García Kerdan, I., Escutia-Guadarrama, L., Reyna-González, J. M., Magaña, J. J., & Leyva-Gómez, G. (2021). Nanoremediation: Nanomaterials and nanotechnologies for environmental cleanup. *Frontiers in Environmental Science*, *9*, 793765. https://doi.org/10.3389/fenvs.2021.793765

Desmond, L. J., Phan, A. N., & Gentile, P. (2021). Critical overview on the green synthesis of carbon quantum dots and their application for cancer therapy. *Environmental Science: Nano*, *8*(4), 848–862. https://doi.org/10.1039/D1EN00017A

Ebrahim, A., Kandasamy, S., Isleem, H. F., Khan, A. H., Prasad, C., Mohammed, A. S., & Alyaseen, A. (2024). Synergistic effects of combined multi-walled carbon nanotubes and glass fibers on concrete: Experimental and economic analysis. *Fullerenes, Nanotubes and Carbon Nanostructures*, 1–17. https://doi.org/10.1080/1536383X.2024.2445668

Gottardo, S., Mech, A., Drbohlavová, J., Małyska, A., Bøwadt, S., Sintes, J. R., & Rauscher, H. (2021). Towards safe and sustainable innovation in nanotechnology: State-of-play for smart nanomaterials. *NanoImpact*, *21*, 100297. https://doi.org/10.1016/j.impact.2021.100297

Harris, M., Hussain, T., Mohamed, H. I., Khan, A., Ansari, M. S., Tauseef, A., . . . & Akhtar, N. (2023). Nanotechnology–A new frontier of nano-farming in agricultural and food production and its development. *Science of the Total Environment*, *857*, 159639. https://doi.org/10.1016/j.scitotenv.2022.159639

Jagtiani, E. (2022). Advancements in nanotechnology for food science and industry. *Food Frontiers*, *3*(1), 56–82. https://doi.org/10.1002/fft2.104

Kiran, B. R., Kopperi, H., & Venkata Mohan, S. (2022). Micro/nano-plastics occurrence, identification, risk analysis, and mitigation: Challenges and perspectives. *Reviews in environmental science and bio/technology*, *21*(1), 169–203. https://doi.org/10.1007/s11157-021-09609-6

Kong, X. J., & Li, J. R. (2021). An overview of metal-organic frameworks for green chemical engineering. *Engineering*, *7*(8), 1115–1139. https://doi.org/10.1016/j.eng.2021.07.001

Kuhn, R., Bryant, I. M., Jensch, R., & Böllmann, J. (2022). Applications of environmental nanotechnologies in remediation, wastewater treatment, drinking water treatment, and agriculture. *Applied Nano*, *3*(1), 54–90. https://doi.org/10.3390/applnano3010005

Kumar, A., Jayeoye, T. J., Mohite, P., Singh, S., Rajput, T., Munde, S., . . . & Parihar, A. (2024). Sustainable and consumer-centric nanotechnology-based materials: An update on the multifaceted applications, risks, and tremendous opportunities. *Nano-Structures & Nano-Objects*, *38*, 101148. https://doi.org/10.1016/j.nanoso.2024.101148

Kumar, R., Shafiq, N., Kumar, A., & Jhatial, A. A. (2021). Investigating embodied carbon, mechanical properties, and durability of high-performance concrete using ternary and quaternary blends of metakaolin, nano-silica, and fly ash. *Environmental Science and Pollution Research*, *28*, 49074–49088. https://doi.org/10.1007/s11356-021-13918-2

Kumar, S. (2023). Smart and innovative nanotechnology applications for water purification. *Hybrid Advances*, *3*, 100044. https://doi.org/10.1016/j.hybadv.2023.100044

Kuperkar, K., Atanase, L. I., Bahadur, A., Crivei, I. C., & Bahadur, P. (2024). Degradable polymeric bio (nano) materials and their biomedical applications: A comprehensive overview and recent updates. *Polymers*, *16*(2), 206. https://doi.org/10.3390/polym16020206

Li, C., Ahmad, S. F., Ayassrah, A. Y. B. A., Irshad, M., Telba, A. A., Awwad, E. M., & Majid, M. I. (2023). Green production and green technology for sustainability: The mediating role of waste reduction and energy use. *Heliyon*, *9*(12). https://doi.org/10.1016/j.heliyon.2023.e22496

Liu, X., Sathishkumar, K., Zhang, H., Saxena, K. K., Zhang, F., Naraginiti, S., . . . & Guo, X. (2024). Frontiers in environmental cleanup: Recent advances in remediation of emerging pollutants from soil and water. *Journal of Hazardous Materials Advances*, 100461. https://doi.org/10.1016/j.hazadv.2024.100461

Liu, Z., Wu, Z., Yao, Q., Cao, Y., Chai, O. J. H., & Xie, J. (2021). Correlations between the fundamentals and applications of ultrasmall metal nanoclusters: Recent advances in catalysis and biomedical applications. *Nano Today*, *36*, 101053. https://doi.org/10.1016/j.nantod.2020.101053

Modi, S., Prajapati, R., Inwati, G. K., Deepa, N., Tirth, V., Yadav, V. K., . . . & Jeon, B. H. (2021). Recent trends in fascinating applications of nanotechnology in allied health sciences. *Crystals*, *12*(1), 39. https://doi.org/10.3390/cryst12010039

Nair, G. M., Sajini, T., & Mathew, B. (2022). Advanced green approaches for metal and metal oxide nanoparticles synthesis and their environmental applications. *Talanta Open*, *5*, 100080. https://doi.org/10.1016/j.talo.2021.100080

Okem, E. S., Iluyomade, T. D., & Akande, D. O. (2024). Revolutionizing US Pavement Infrastructure: A pathway to sustainability and resilience through nanotechnology and AI Innovations. *World Journal of Advanced Engineering Technology and Sciences*, *11*(2), 411–428. https://doi.org/10.30574/wjaets.2024.11.2.0124

Popescu, M., & Ungureanu, C. (2023). Green nanomaterials for smart textiles are dedicated to environmental and biomedical applications. *Materials*, *16*(11), 4075. https://doi.org/10.3390/ma16114075

Reents, T., Kataev, E., Duarte-Ruiz, D., Wilks, R. G., Garcia-Diez, R., Bär, M., & Cocchi, C. (2025). Combinatorial ab initio calculations and core spectroscopy unravel the electronic structure of nickel cobalt manganese oxide. *Scientific Reports*, *15*(1), 5816. https://doi.org/10.1038/s41598-025-89283-8

Roy, A., Sharma, A., Yadav, S., Jule, L. T., & Krishnaraj, R. (2021). Nanomaterials for remediation of environmental pollutants. *Bioinorganic Chemistry and Applications*, *2021*(1), 1764647. https://doi.org/10.1155/2021/1764647

Sindhwani, S., & Chan, W. C. (2021). Nanotechnology for modern medicine: Next step towards clinical translation. *Journal of Internal Medicine*, *290*(3), 486–498. https://doi.org/10.1111/joim.13254

Winiarska, E., Jutel, M., & Zemelka-Wiacek, M. (2024). The potential impact of nano-and microplastics on human health: Understanding human health risks. *Environmental Research*, 118535. https://doi.org/10.1016/j.envres.2024.118535

Zhang, P., Guo, Z., Ullah, S., Melagraki, G., Afantitis, A., & Lynch, I. (2021). Nanotechnology and artificial intelligence to enable sustainable and precision agriculture. *Nature Plants*, *7*(7), 864–876. https://doi.org/10.1038/s41477-021-00946-6

2 Self-Healing Polymers
Formulation and Properties

Vijay Chaudhary, Harshu Kumar, and Chanchal Ahlawat

2.1 INTRODUCTION

A polymer is a macromolecule made up of many small repeating sub-units, also known as monomers, which are joined by covalent bonds. Polymers can be broadly classified into two types based on the nature of their origin, namely natural and synthetic polymer. Examples of natural polymers are DNA, protein, skin tissue, hair, and cellulose, and examples of synthetic polymers are rubber (elastomers), plastic, all types of adhesives, PVC, polypropylene (PP), and resins [1]. These days polymers play a vital role in our day-to-day life for a variety of purposes. For example, plastic is one of the most commonly used polymers that can be found everywhere from house to industry in the form of PET bottles, biodegradable plastic bags, plastic boxes, etc. Another way to classify polymer is based on molecular forces, which are thermoplastic and thermoset polymers. Thermoplastic polymers can be reshaped upon heating and are not permanently cross-linked. They can sustain multiple cycles of reshape upon exposure of heat with negligible degradation; hence, they are more recyclable in nature. On the other hand, thermoset polymers are cross-linked (cured) upon addition of energy (typically in the form of heat). They can't be reshaped upon exposure of heat or any other energy because during curing process the macromolecules get permanently cross-linked and included in the molecular network structure. They are stiffer and stronger in nature than thermoplastic polymers [2]. Examples of thermoplastic polymers are polypropylene (PP), polyvinyl chloride (PVC), low-density polyethylene, and high-density polyethylene, and examples of thermoset polymers are bakelite, epoxy resins, polyester, polyurethane, and melamine resins [3].

Self-healing polymers are another specific type of polymers that have the special ability to repair themselves without any external interference, which brings them to their initial state of properties after damage. Due to their special ability, there is wide of range of applications in electronics, space industry, biomedical industry, protective environment, etc. [4]. They also possess applications beyond conventional mechanical strength such as in food packaging, radiation shielding, tissue regeneration, and acoustic conservation [5]. Self-healing polymers can be broadly classified into two categories based on their self-healing mechanism which are intrinsic self-healing and extrinsic self-healing. Extrinsic self-healing has an immersed healing agent inside their polymer matrix via microcapsule or vascular networks. When damage occurs,

DOI: 10.1201/9781003639510-2

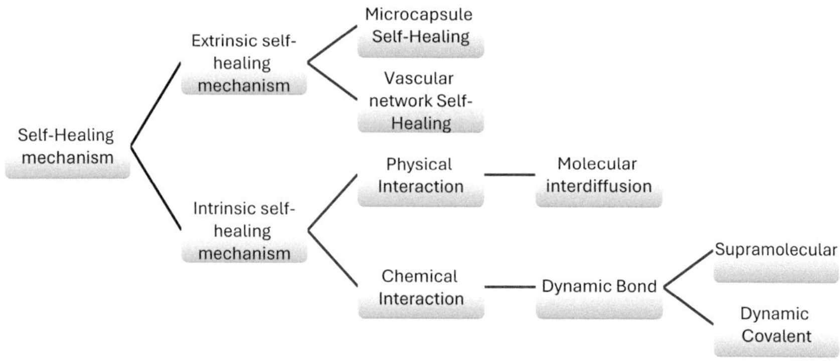

FIGURE 2.1 Classification of self-healing mechanism.

the matrix breaks, which releases the healing agent that reacts with each other or reacts with the matrix to repair the cracked or damaged surface. It has ample of applications where automatic healing is required without any external intervention. However, its usage is constrained by the consumption of healing agent, thereby limiting multiple healing at the same site of the material [4]. Whereas, intrinsic self-healing works a bit differently, as they are independent of healing agent consumption because their healing phenomenon is based on dynamic bonding that gets rapidly developed and is repetitive in nature. However, the challenge in dealing with intrinsic healing is the trade-off between the mechanical strength and its self-healing performance, where both qualities can't be increased simultaneously. Mainly this phenomenon occurs due to the fundamental differences in the polymeric motion. Although stronger interactions between polymer chains enhance mechanical strength, they also restrict chain mobility, which in turn reduces self-healing performance. [6, 7]. Figure 2.1 represents the classification of self-healing mechanism.

2.2 CLASSIFICATION OF SELF-HEALING MECHANISM

2.2.1 Extrinsic Self-Healing Mechanism

Extrinsic self-healing incorporates the involvement of heterogeneous materials into the polymer matrix to fill-up the area where cracks have been formed. There are several types of mechanisms that fall into the extrinsic type of self-healing, which are in the form of microcapsule self-healing and vascular network self-healing.

2.2.1.1 Capsular Type Self-Healing

This type of self-healing utilizes the healing agent stored in microcapsules present inside the matrix, which are released towards the cracked site via capillary effect. Also, when the healing agent gets in contact with the catalyst, which spreads throughout the matrix, it initiates the cross-linking or polymerization that ultimately restores the damaged space [6]. White et al. developed microcapsules using urea formaldehyde, which are of 50–200 micrometres in size and are filled by

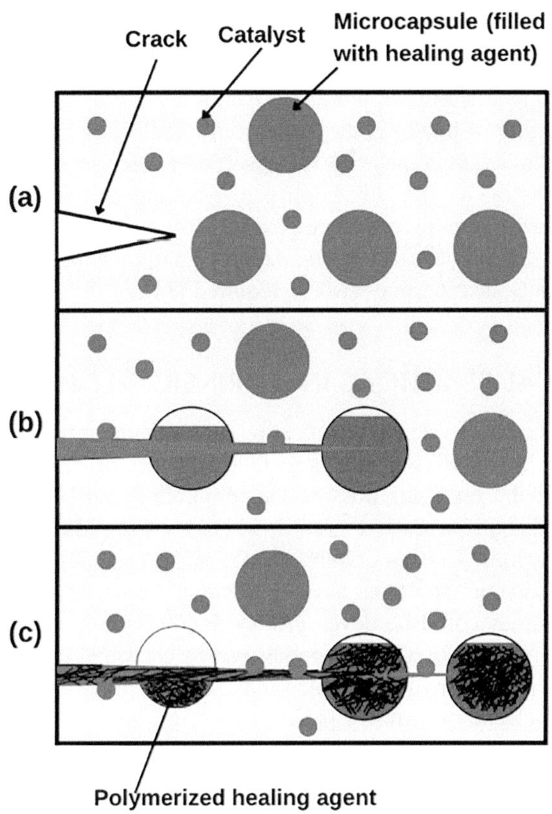

FIGURE 2.2 Microcapsule self-healing process.

dicyclopentadiene (DCPD), which act as the healing agent. These DCPD-filled capsules are immersed throughout the matrix along with the Grubb's catalyst. During usage, cracks formed in the matrix get propagated and break the microcapsules, after which the DCPD (healing agent) gets released and interacts with the Grubb's catalyst, due to which polymerization is initiated and cracks get filled [8]. The process is illustrated in Figure 2.2.

2.2.1.2 Vascular Type Self-Healing

This type of self-healing mimics the biological vascular system by incorporation of interconnected microchannels that deliver the healing agent to the damaged sites. The healing agent is being stored in the microchannels of the vascular networks like blood vessels as in human beings. The functioning of the vascular type is similar to the capsular type; when a polymer is damaged upon crack propagation, if a microchannel is ruptured, it releases the healing agent, which when comes in contact with the catalyst initiates polymerization and the cracks get filled [9]. This self-healing type allows continuous or multiple healing cycles at the same site by connecting one of the interconnected channels with the external reservoir of healing agent [10].

2.2.2 Intrinsic Self-Healing

Intrinsic self-healing mechanism restores the damage site by means of dynamic bonding/inherent reversibility of bonding capabilities of polymer matrix. Due to this inherent reversibility of bonding, they offer multiple healing cycles at the same site without the need of external material. However, they can only be utilized properly at micro-level damages, as the reversibility of bonding can occur only when the chemical's distances are in nanometre range. Therefore, to initiate the healing mechanism, it is necessary to bring the damaged parts together, and while dealing with hard materials, heating may also be required [11].

2.3 CHEMICAL REACTIONS IN INTRINSIC SELF-HEALING

In intrinsic self-healing systems, chemical reaction occurs at the damage site, while the reversibility of bonding is the key factor governing the healing process. . However not all the reactions possess the re-bonding capabilities. Some investigations suggest that a few reactive groups, which are free radicals, hydrogen bonds, ionomeric interactions, C=C, –COOH, –NH2, –OH, –SH, –Si–O, S–S, C=O, and the formation of cyclic structures have self-healing capabilities, and various other self-healing materials could be developed by the incorporation of the above-mentioned reactive groups [12]. These materials can be classified based on the nature of the bonding mechanism involved, which may be reversible covalent bond or those based on supramolecular chemistry [13].

Reversible covalent bonds: These bonds possess the unique ability to break and form chemical bonds between different molecules or functional groups within a polymer. This reversibility of bond brings equilibrium to molecules and enables the initiation of self-healing process [14]. Polymers that have these reversible covalent bonds can rearrange their structures and bonding patterns through these reversible reactions, which are initiated by external stimuli such as heat, variation in pH, and light. After the removal of these stimuli, polymers regain their original stability and bonding strength similar to irreversible covalent polymers [15]. Therefore, the reversible covalent bonds exhibit better repeatable repairing of polymers while maintaining the strong mechanical properties. However, the one significant drawback with these reversible covalent bonds is that they require high activation energy, due to which a larger amount of energy (in the form of light or heat) is being involved in breaking and forming of bonds [16].

Supramolecular chemistry: In contrast to reversible covalent bonds, supramolecular interactions offer rapid remodelling of polymer network, which allows the materials to transition quickly from fluid-like state with low density and high free volumes to more solid-like, elastic, and plastic networks with lower free volume. However, these systems have certain limitations. Polymers typically based on supramolecular chemistry have quite lower glass transition temperature, inferior mechanical properties, and due to their uses are restricted to some specific application. The self-healing properties of supramolecular materials arise due to the formation of non-covalent interactions such as hydrogen bonds, metal coordination, π–π stacking, ionic interactions, and host–guest interactions [13, 17, 18]. Upon the application of mechanical forces, the weak supramolecular bonds dissociate. However, due to the reversible nature of these bonds, they could be easily reformed enabling the material to regenerate and re-heal itself [13].

2.4 PHYSICAL INTERACTION IN INTRINSIC SELF-HEALING

In intrinsic self-healing, the healing process can be achieved also by physical interaction, where the damage site gets repaired upon by the entropy of molecular diffusion, which is further initiated by external stimuli such as heating [19]. While heating, the polymer chain gets diffused or gets mobilized. Due to this, the mobility of the chains allows them to diffuse across damaged regions within the material. This allows the formation of new entanglements with neighbouring chains, which helps to close and heal the crack [20, 21]. This process is usually effective when the temperature reaches above the glass transition temperature of the polymer or the melting temperature, as chain diffusions and re-entanglement occur more prominently in these conditions [22].

Thermoplastic polymer and thermosetting polymer both possess the capability of self-healing by molecular diffusion. In case of thermoplastic, the molecular diffusions get started as the temperature gets above its glass transition temperature and after which the polymer mobility occurs. When the damaged parts are brought in contact, due to these diffusions or polymer mobility, cracks are healed, thereby restoring the material's mechanical strength [23]. It has been observed that the above healing mechanism in thermoplastic is generally at a temperature range between −50 and 100°C at atmospheric pressure or in vacuum [24].

In case of thermosetting polymers, healing through the physical interaction can be achieved at ambient or increased temperature where healing of cracks can be done through interdiffusion of dangling chains. But thermosetting polymer has relatively more limitations than thermoplastic due to the presence of cross-link chains, which form an irreversible chemical bond [25]. Therefore, these polymers have high glass transition temperature, and the mobility of polymer chains also occurs at a higher temperature. This may result in molecular degradation, due to which, the material could also lose its stability [19]. However, this limitation could be minimized by impregnation of thermoplastic into thermosets polymer matrix, where healing can be caused by first heating, which helps in mobilizing the impregnated thermoplastic only towards the damage site due to which cracks get filled and the original thermosets remain unaffected or stable [26].

2.5 SELF-HEALING IN POLYMER COMPOSITES

2.5.1 Carbon Fibre-Reinforced Self-Healing Polymer Composites

Kessler et al. [27] demonstrated a self-healing fibre-reinforced composite in which a microencapsulated healing agent along with a solid chemical catalyst has been spread throughout the polymer matrix. Here the healing agent is dicyclopentadiene (DCPD) monomer, which has low viscosity and excellent shell life. While stabilizing with 100–200 ppm of p-tert-butylcatechol, the catalyst used is bis (tricyclohexylpho sphine) benzylidine ruthenium (IV) dichloride also known as Grubbs' catalyst. An epoxy-based polymer matrix is being used and carbon fibre is used as reinforcement. Upon facing the damage into the composite, the cracks propagate and rupture the microcapsules, which releases the healing agent into the crack plane, where it meets the catalyst and initiates the polymerization and subsequently the bonding of crack faces takes place. While it has been found out that the healing at room temperature can yield 45% recovery of virgin interlaminar fracture toughness, at 80°C up to 80% of recovery has been found out [27].

2.5.2 Glass Fibre-Reinforced Self-Healing Polymer Composites

Trask et al. [28] consider the storage of healing material component inside the hollow glass fibres (HGF) and placement of this HGF piles within the glass fibre/ epoxy and carbon fibre/ epoxy laminates to heal the damage and restore their mechanical strength. Their study suggests that inclusion of HGF initially reduces strength but enhances damage tolerance.

These composites also show significant recovery of flexural strength through self-healing [28]. Another glass fibre-reinforced self-healing polymer composite fabricated using urea-formaldehyde microcapsules contains epoxy resin and hardener. The composition of microcapsules is about 10% by wt. in the composite incorporated along with the glass fibre and epoxy resin. The results show that there was 42% improvement in impact resistance after healing and 17% increase in tensile strength when compared with the damaged state [29].

2.5.3 Aramid Fibre-Reinforced Self-Healing Composites

Sun et al. [30] focussed on the problems of soft composites with high puncture resistance and self-healing at ambient temperature. The author synthesized a room temperature self-healing thermoplastic polyurethane (TPU) coating through phase modulation and tried to fabricate a type of puncture resistance composite with high performance aramid fabrics and compost it with the TPU. The self-healing TPU made a stronger fibre matrix interface through intermolecular H-bonding with the aramid fabrics. The results of the experiment show that the maximum spike-puncture force and maximum knife-stab force of the composite were 14.69 and 4.22 times higher than those of pure aramid fabrics, respectively. Also, the healing efficiency of the spike stab resistance is as high as 44.91% and the self-healing efficiency of the knife stab is as high as 66.98% after 48 h at 30°C [30].

2.5.4 Self-Healing Concrete Composites

2.5.4.1 Microcapsule-Based Self-Healing Concrete Composites

A study was conducted on the potential use of self-healing in concrete where microencapsulated epoxy resin was used as a healing agent. Various effects have been encountered while incorporating the healing agent such as minor reduction in concrete properties like slump and density and minor increase in air content. However, the mechanical properties of this self-healing concrete were found to be fully restored while simultaneously healing cracks, with compressive strength increased from 38 to 40 MPa (same as in initial state) after healing, flexural strength restored to 5 MPa after repairing, and slight improvement in tensile strength (3.3 MPa) from its initial strength (3.1 MPa). Despite some minor trade-off in concrete properties, significant recovery of post-healing mechanical properties shows that inclusion of self-healing microcapsule can result in ample durability and longevity of concrete structures in practical scenarios [31].

2.5.4.2 Vascular Self-Healing Concrete

A study investigated the introduction of 3-D vascular tube network for a self-healing concrete by performing several tests and monitoring its healing and mechanical recovery. The tubes fabricated by 3-D printing were found to be as effective as traditional ceramic tubes, and they break instantly during of crack formation as the healing agent releases instantly. Healed cracks show 80% recovery in strength and toughness compared to their original state. However, there are several challenges: the bulk reservoir of healing agent in the concrete weakens its resistance to failure; challenges related to design, such as the clogging of vascular tubes as the healing agents are released, making next healing cycle impossible; and the vacuum formation under the tubes during casting leads to limited generation of vibration, which can affect concrete strength [32, 33].

2.5.5 Self-Healing Polymer-Based Metal Matrix Composites

2.5.5.1 Aluminium-Silicon (Al-Si) Matrix Composites

It is one of the NASA's flagship composites reinforced by 2% by vol. of Ni Ti shape memory alloy (SMA) wires with aluminium-silicon matrix. It has a healing efficiency of 91.6% under tensile conditions. It can operate in a dual-step healing mechanism where healing above the SMA's austenite start temperature (approx. 400°C) initiates both crack closure through SMA contractions and matrix liquefaction through eutectic phase melting. It has been tested extensively for fatigue crack repair, and it demonstrates the ability to heal large fatigue cracks that occur during service life. The test results shows that it has an ability to recover more than 90% of its original ultimate tensile strength value after healing cycles, making it a promising product for its application in aerospace structural components such as fuselage skin, frames, ribs, and fuel lines [34, 35].

2.5.5.2 Titanium-Boride (Ti-TiB) Composites

Titanium-boride composites have been fabricated specifically for their focus on automobile applications, such as for engine valves mass production for passenger vehicles. This composite consists of particles of titanium-boride with titanium matrix along with self-healing capabilities through liquid-assist healing mechanism, combined with shape memory effects [36].

2.5.5.3 Nickel-Titanium Superalloys

This system is developed specifically for high temperature applications where $\eta - Ni_3Ti$ precipitates are being used as healing agent. These healing Ni-alloys found remarkable improvement in performance where the creep lifetimes are found to be 8-10 times higher than the non-self-healing variants. The healing mechanism operates via the defect-site-specific precipitation, and Ti atoms diffuse towards the damaged sites and form the healing precipitates at temperatures around 750°C [37].

2.5.5.4 Tin-Bismuth (Sn-Bi) Matrix Composites

This system is considered as one of the effective self-healing systems. Under research conducted by University of Wisconsin-Milwaukee, this composite was developed using eutectic Sn-Bi alloy matrix, which is reinforced by NiTi SMA wires. This composite achieves its self-healing ability mainly through two mechanisms. One of them is 'shape memory effect' where the NiTi wires get back to their original geometry by reverting to their parent shape when they are being heated beyond their austenite transition temperature. And other mechanism is 'liquid-assist healing' where the eutectic material in the matrix melts at healing temperature (165° C), flows into the cracks, and solidifies to restore its structural integrity. Testing results show its strength recovery of upto 92%, with the healing process completed in 74 seconds of thermal activation [38].

2.6 FACTORS AFFECTING SELF-HEALING

Table 2.1 gives a detailed comparison of factors affecting self-healing mechanism in both intrinsic and extrinsic type self-healing.

TABLE 2.1
Showing the Factors Affecting Self-Healing for Intrinsic and Extrinsic Self-Healing

S. No.	Factors Affecting Self-Healing	Intrinsic Self-Healing	Extrinsic Self-Healing
1.	**Healing mechanism**	Reversible covalent bonds or supramolecular interactions within the material's own structure	Release of healing agents from embedded capsules, fibres, or vascular networks upon damage
2.	**Stimulus**	Damage triggers molecular rearrangement; can also be activated by heat, light, solvents, or mechanical stress.	Damage ruptures capsules/ networks, releasing healing agent; can also be triggered by heat or other stimuli.
3.	**Healing agent**	No need for external healing agent	Healing agent is stored separately and delivered to the damage site
4.	**No. of healing cycle possibility**	Multiple healing cycles is possible as healing based on reversible reaction.	Typically limited to the amount of healing agent available; once depleted, healing stops.
5.	**Healing Efficiency**	Depends upon chain mobility, bond dynamics, and environmental conditions.	Can exceed 100% for some properties; dependent on agent delivery and reaction efficiency.

(Continued)

TABLE 2.1 (Continued)

S. No.	Factors Affecting Self-Healing	Intrinsic Self-Healing	Extrinsic Self-Healing
6.	**Mechanical Properties**	Often lower due to need for chain mobility and weaker dynamic bonds.	Can be designed for higher strength, but may be compromised by agent incorporation.
7.	**Healing Time**	Depends on molecular mobility and temperature; faster at higher temperatures.	Depends on agent transport and reaction kinetics.
8.	**Examples**	Epoxies, polyacrylates, polyamides with DA reactions, supramolecular polymers.	Microcapsule- or fibre-based composites, vascular networks in polymers.

2.7 INDUSTRY-SPECIFIC APPLICATIONS OF SELF-HEALING MATERIALS

Self-healing materials are proved to be a groundbreaking solution for various applications due to their ability to repair damage of the products, extend their life cycles, and subsequently reduce the cost related to maintenance and other costs. Various studies have already explored its applications, largely driven by its mechanical properties and durability, particularly in materials science as well as in housing and protective coatings for electronics. However, there is still need for exploring their industry-specific usage in various industries like aerospace, energy storage, biotechnology, construction and infrastructure, and automobiles [39].

2.7.1 Self-Healing in Aerospace Industry

In the manufacturing of fuselage in wide-body aircraft like Boeing 787 and Airbus A350 it has been found that upto 50% (by wt.) of its composition contains fibre reinforced polymer (FRP) composites. However, the currently used FRPs are very sensitive towards damages due impact load. To address this issue there might be need to consider a heavier design for safety concerns, which leads to the increase in weight of the overall aircraft. Additionally, detecting such impact damage can be challenging during maintenance. Self-healing can be one of the effective solutions to this problem by offering lesser vulnerability towards impact load and a light weight and low- maintenance solution [40]. A study was carried out on a self-healing hollow glass fibre-epoxy composite by Tan et al., [41] where he performed three-point bend impact tests on this composite, and the post-damage flexural test results show upto 47% strength recovery of composite after healing, which ultimately improved the impact damage resistance capability of aircraft [41].

Another study conducted by Guadagno et al. [42] related to the effect of healing components on the mechanical properties of composites for aerospace structural application. It focussed on the composite with DGEBA (diglycidyl ether of bisphenol A) epoxy matrix embedded with DCPD-containing urea-formaldehyde microcapsules. Due to the addition of catalyst in the matrix, the elastic modulus was found to be slightly decreased. However, there was significant recovery of virgin mechanical properties after healing [42].

Xiang et al. [43] developed a hybrid carbon fibre/epoxy composite featuring self-healing core-shell nanofibres at the interfaces, utilizing a co-electrospinning technique. In their design, dicyclopentadiene (DCPD) was encapsulated within polyacrylonitrile (PAN) to form core-shell DCPD/PAN nanofibres, with DCPD serving as the healing agent. This innovative approach demonstrates that lightweight polymer matrix can be employed in aerospace applications, providing both high strength and high toughness. The self-healing capability helps reduce damage and restores the functionality and service life of affected components. Polymer matrix composites are particularly attractive for aerospace use due to their ease of manufacturing and cost-effectiveness [43].

2.7.2 Self-Healing in Automotive Industry

A long durable tyre with lesser fuel consumption without compromising safety standards is quite desirable and demanding in automotive industry. To address the same, a self-healing tyre has been developed by a researcher with the use of styrene butadiene rubber matrix with reclaimed tyre waste as a sustainable filler. A further addition of ground tyre rubber into matrix provides better concurrent rolling resistance, wet grip, and mending ability. The researcher found in his recent study that it has gained 100% recovery in stiffness and relaxation dynamic because of cyclic formation, which in turn leads to the formation of heterogeneous rubber network [44].

2.7.3 Self-Healing in Construction Industry

Concrete is one of the smallest and most fundamental components of the construction industry and is considered as the backbone of this industry. It is primarily used for construction of buildings due to its properties like high compressive strength, favourable fire resistance, ease of casting, and cost-effectiveness. However, due its low tensile strength, it is quite susceptible to damages upon breaking [45]. The use of self-healing could offer a promising solution to minimize concrete damages and cracking by filling the cracks with the healing agent along with a suitable catalyst [46]. In 2017, A. Kanellopoulos investigated the use of bacterial-based agents to synthesize microcapsules made of gelatine and acacia gum for encapsulating sodium silicate liquid in self-healing concrete and found the approach to be highly effective and well-balanced [47]. Self-healing agents such as superabsorbent polymers (SAP) and polyurethane encapsulation materials improve the water tightness of concrete, making them suitable for structures exposed to water, like tunnels, dams, and basements [48]. Self-healing polymers incorporated into asphalt binders can repair micro-cracks caused by traffic loads, improving the durability and performance of road surfaces [49]. Table 2.2 shows some of the key self-healing agents in construction industry with their specific use and key benefits.

TABLE 2.2
Different Self-Healing Agents with Their Specific Use and Key Benefits

Self-Healing Agent/ Method	Application Area	Key Benefits
Microcapsules, SAP, bacteria	Concrete structures	Crack repair, durability, water resistance
Self-healing polymers in binder	Road construction (asphalt)	Fatigue life, reduced maintenance
Bacteria in expanded perlite	Marine concrete	Crack closure in harsh environments
Polyurethane encapsulation	Cementitious materials	Improved water tightness, efficient crack repair

2.7.4 Self-Healing in Space Industry

In space, the integrity of spacecraft is threatened not only by micrometeoroids and orbital debris but also by extreme temperature fluctuations. To address these challenges, self-healing composites have emerged as a promising solution. These advanced materials offer both autonomous repair capabilities and effective thermal protection. As a result, they are highly suitable for manufacturing circuits, rockets, and other space mechanisms that demand the highest safety standards [50]. In 2019, Zhu et al. [51] introduced UV-responsive self-healing microcapsules for aerospace coatings. They developed a coating system that utilizes these microcapsules to enable in-orbit damage repair when exposed to ultraviolet light. As a result, coatings have become an essential method in the space industry, serving as protective barriers for critical components such as fuselages, wings, and engines [51].

2.8 CONCLUSION

This review investigates the self-healing polymers that possess self-healing or autonomous repairing capabilities without any external interference. Due to this unique property of this type of polymers, there is an abundance of applications of self-healing materials in various industries like aerospace, spacecraft, automobile, and construction. This chapter also dives deep into the classification of self-healing, which are extrinsic and intrinsic self-healing. Extrinsic self-healing is dependent on self-healing agents that are stored in microcapsules, in vascular network forms (like the blood flow in blood vessels in biological bodies), and various other forms. The healing cycles are highly dependent on healing agent reservoir and therefore multiple healing cycles are not possible at the same site by extrinsic self-healing. However, they are mostly known for autonomous repairing without any external stimuli, hence they have applications where repairing doesn't require any external intervention.

On the other hand, intrinsically self-healing mechanisms are quite different from extrinsic self-healing as they leverage the use of reversible chemical reactions like reverse covalent bonds, Diels-Alder reactions, disulfide exchange, and supramolecular interaction (such as hydrogen bonding). While intrinsic systems are highly promising, they often require a careful balance between mechanical strength and healing efficiency, as increased cross-linking can restrict polymer chain mobility and thus limit self-healing performance. Similarly, extrinsic self-healing can be done in microcapsular form and vascular form, wherein capsular form healing is quite restricted due to limited healing agent stock, but the vascular network form can be refilled with extra healing agent as per requirements, as it can be connected to the external reservoirs with one of its channel or network.

The application of self-healing polymers in composite materials, such as those reinforced with carbon, glass, or aramid fibres, has opened new avenues in aerospace, civil engineering, electronics, and protective coatings. These composites not only enhance the mechanical robustness of structures but also endow them with the ability to recover from micro-damages, thereby extending their service life and reliability. Notably, the integration of self-healing functionalities in concrete and fibre-reinforced composites has shown substantial improvements in impact resistance, tensile strength, and puncture resistance, underscoring their potential for structural and safety-critical applications.

Despite the impressive progress, challenges remain. Achieving an optimal balance between mechanical properties and healing capability, scaling up manufacturing processes, and ensuring long-term stability and repeatability are areas that require further research. Additionally, the development of self-healing systems that can operate effectively under a wide range of environmental conditions—including extreme temperatures and radiation—will be essential for broader adoption in demanding fields such as aerospace and space exploration.

To sum-up the above, it can be said that self-healing polymers and their composites hold immense promise for the future of smart materials. Their ability to autonomously repair damage not only reduces maintenance costs and downtime but also enhances the durability and safety of critical structures. Continued innovation and interdisciplinary collaboration will be key to overcoming existing limitations and unlocking the full potential of self-healing materials across diverse industries.

REFERENCES

[1] H. Baskan-Bayrak, N. Aliyeva, M. S. Sorayani Bafqi, and B. S. Okan, "Classification of waste plastics for dimension-controlled graphene growth on natural mineral substrates in terms of polymer processing and thermal techniques," *Graphene Extraction from Waste: A Sustainable Synthesis Approach for Graphene and Its Derivatives*, edited by Chaudhery Mustansar Hussain, Ajeet Kumar Srivastav, . . . James M. Tour, pp. 117–149, Jan. 2023, doi:10.1016/B978-0-323-90914-3.00012-7

[2] J. Bergström, "Introduction and overview," *Mech Solid Polym*, pp. 1–17, Jan. 2015, doi:10.1016/B978-0-323-31150-2.00001-7

[3] A. F. Ahmad, A. R. Razali, and I. S. M. Razelan, "Utilization of polyethylene terephthalate (PET) in asphalt pavement: A review," *IOP Conf Ser Mater Sci Eng*, vol. 203, no. 1, Jun. 2017, doi:10.1088/1757-899X/203/1/012004

[4] D. Y. Wu, S. Meure, and D. Solomon, "Self-healing polymeric materials: A review of recent developments," *Prog Polym Sci (Oxford)*, vol. 33, no. 5, pp. 479–522, May 2008, doi:10.1016/j.progpolymsci.2008.02.001

[5] S. Kim, H. Jeon, J. M. Koo, D. X. Oh, and J. Park, "Practical applications of self-healing polymers beyond mechanical and electrical recovery," *Adv Sci*, vol. 11, no. 16, Apr. 2024, doi:10.1002/ADVS.202302463
[6] K. Choi, A. Noh, J. Kim, P. H. Hong, M. J. Ko, and S. W. Hong, "Properties and applications of self-healing polymeric materials: A review," *Polymers*, vol. 15, no. 22, p. 4408, Nov. 2023, doi:10.3390/POLYM15224408
[7] J. Luo, T. Wang, C. Sim, and Y. Li, "Mini-review of self-healing mechanism and formulation optimization of polyurea coating," *Polymers*, vol. 14, no. 14, p. 2808, Jul. 2022, doi:10.3390/POLYM14142808
[8] S. R. White et al., "Autonomic healing of polymer composites," *Nature*, vol. 409, no. 6822, pp. 794–797, Feb. 2001, https://www.nature.com/articles/35057232
[9] H. R. Williams, R. S. Trask, A. C. Knights, E. R. Williams, and I. P. Bond, "Biomimetic reliability strategies for self-healing vascular networks in engineering materials," *J R Soc Interface*, vol. 5, no. 24, pp. 735–747, Jul. 2008, https://doi.org/10.1098/rsif.2007.1251
[10] T. C. Mauldin and M. R. Kessler, "Self-healing polymers and composites," *Int Mater Rev*, vol. 55, no. 6, pp. 317–346, Nov. 2010, doi:10.1179/095066010X12646898728408
[11] C. E. Diesendruck, N. R. Sottos, J. S. Moore, and S. R. White, "Biomimetic self-healing," *Angew Chem Int Ed*, vol. 54, no. 36, pp. 10428–10447, Sep. 2015, doi:10.1002/anie.201500484
[12] Y. Yang and M. W. Urban, "Self-healing polymeric materials," *Chem Soc Rev*, vol. 42, no. 17, p. 7446, 2013, doi:10.1039/c3cs60109a
[13] Y. Yang, X. Ding, and M. W. Urban, "Chemical and physical aspects of self-healing materials," *Prog Polym Sci*, vol. 49–50, pp. 34–59, Oct. 2015, doi:10.1016/j.progpolymsci.2015.06.001
[14] J. Dahlke, S. Zechel, M. D. Hager, and U. S. Schubert, "How to design a self-healing polymer: General concepts of dynamic covalent bonds and their application for intrinsic healable materials," *Adv Mater Interfaces*, vol. 5, no. 17, Sep. 2018, doi:10.1002/admi.201800051
[15] Z. P. Zhang, M. Z. Rong, and M. Q. Zhang, "Polymer engineering based on reversible covalent chemistry: A promising innovative pathway towards new materials and new functionalities," *Prog Polym Sci*, vol. 80, pp. 39–93, May 2018, doi:10.1016/j.progpolymsci.2018.03.002
[16] J. A. Syrett, C. R. Becer, and D. M. Haddleton, "Self-healing and self-mendable polymers," *Polym Chem*, vol. 1, no. 7, p. 978, 2010, doi:10.1039/c0py00104j
[17] J.-M. Lehn, "From supramolecular chemistry towards constitutional dynamic chemistry and adaptive chemistry," *Chem Soc Rev*, vol. 36, no. 2, pp. 151–160, 2007, doi:10.1039/B616752G
[18] J. L. Atwood, *Comprehensive supramolecular chemistry II*. Elsevier, 2017.
[19] K. Urdl, A. Kandelbauer, W. Kern, U. Müller, M. Thebault, and E. Zikulnig-Rusch, "Self-healing of densely crosslinked thermoset polymers—a critical review," *Prog Org Coat*, vol. 104, pp. 232–249, Mar. 2017, doi:10.1016/j.porgcoat.2016.11.010
[20] M. D. Hager, P. Greil, C. Leyens, S. van der Zwaag, and U. S. Schubert, "Self-Healing Materials," *Adv Mater*, vol. 22, no. 47, pp. 5424–5430, Dec. 2010, doi:10.1002/adma.201003036
[21] C. B. Lin, S. Lee, and K. S. Liu, "Methanol-Induced crack healing in poly(methyl methacrylate)," *Polym Eng Sci*, vol. 30, no. 21, pp. 1399–1406, Nov. 1990, doi:10.1002/pen.760302109
[22] S. Bonardd, M. Nandi, J. I. Hernández García, B. Maiti, A. Abramov, and D. Díaz Díaz, "Self-healing polymeric soft actuators," *Chem Rev*, vol. 123, no. 2, pp. 736–810, Jan. 2023, doi:10.1021/ACS.CHEMREV.2C00418/ASSET/IMAGES/MEDIUM/CR2C00418_0047.GIF

[23] S. van der Zwaag, N. H. van Dijk, H. M. Jonkers, S. D. Mookhoek, and W. G. Sloof, "Self-healing behaviour in man-made engineering materials: Bioinspired but taking into account their intrinsic character," *Philos Trans R Soc A Math Phys Eng Sci*, vol. 367, no. 1894, pp. 1689–1704, May 2009, doi:10.1098/rsta.2009.0020
[24] D. Y. Wu, S. Meure, and D. Solomon, "Self-healing polymeric materials: A review of recent developments," *Prog Polym Sci*, vol. 33, no. 5, pp. 479–522, May 2008, doi:10.1016/j.progpolymsci.2008.02.001
[25] V. K. Stokes, "Joining methods for plastics and plastic composites: An overview," *Polym Eng Sci*, vol. 29, no. 19, pp. 1310–1324, Oct. 1989, doi:10.1002/pen.760291903
[26] B. J. Blaiszik, S. L. B. Kramer, S. C. Olugebefola, J. S. Moore, N. R. Sottos, and S. R. White, "Self-Healing polymers and composites," *Annu Rev Mater Res*, vol. 40, no. 1, pp. 179–211, Jun. 2010, doi:10.1146/annurev-matsci-070909-104532
[27] M. R. Kessler, N. R. Sottos, and S. R. White, "Self-healing structural composite materials," doi:10.1016/S1359-835X(03)00138-6
[28] R. S. Trask, G. J. Williams, and I. P. Bond, "Bioinspired self-healing of advanced composite structures using hollow glass fibres," *J R Soc Interface*, vol. 4, no. 13, p. 363, Apr. 2006, doi:10.1098/RSIF.2006.0194
[29] S. N. Kadam et al., "Development and characterization of self-healing glass fiber composites for applications requiring autonomous crack repair," https://www.espublisher.com/, vol. 25, p. 1202, Aug. 2024, doi:10.30919/ESEE1202
[30] Y. Sun et al., "self-healing polyurethane/fabric composite with puncture resistance enabled by phase regulation and matrix-fiber strengthening," *Acs Appl Polym Mater*, vol. 6, no. 20, pp. 12478–12491, Oct. 2024, doi:10.1021/ACSAPM.4C01635/SUPPL_FILE/AP4C01635_SI_001.PDF
[31] V. M. Savadi and R. Santhosh, "Self-healing concrete using microencapsulated epoxy resin: Evaluation of fresh and mechanical properties," *Int J Res Eng Appl Manag (IJREAM)*, vol. 10, p. 7, 2024, doi:10.35291/2454-9150.2024.0402
[32] E. Tsangouri, C. Van Loo, Y. Shields, N. De Belie, K. Van Tittelboom, and D. G. Aggelis, "Reservoir-vascular tubes network for self-healing concrete: Performance analysis by acoustic emission, digital image correlation and ultrasound velocity," *App. Sci (Switzerland)*, vol. 12, no. 10, May 2022, doi:10.3390/app12104821
[33] Y. Shields, V. Cappellesso, T. Van Mullem, N. De Belie, and K. Van Tittelboom, "A comparison of potential healing agents for vascular-based self-healing concrete," doi:10.1051/matecconf/202337802026
[34] C. R. Fisher et al., "Repairing large cracks and reversing fatigue damage in structural metals," *Appl Mater Today*, vol. 13, pp. 64–68, Dec. 2018, doi:10.1016/J.APMT.2018.07.003
[35] M. C. Wright, M. Manuel, and T. Wallace, "Fatigue resistance of liquid-assisted self-repairing aluminum alloys reinforced with shape memory alloys shape memory alloy self-healing (SMASH) technology for aeronautical applications," 2013. Accessed: Jul. 09, 2025. http://www.sti.nasa.gov
[36] H. Ozair, A. H. Baluch, M. A. Ur Rehman, and A. Wadood, "Shape memory hybrid composites," *ACS Omega*, vol. 7, no. 41, p. 36052, Oct. 2022, doi:10.1021/ACSOMEGA.2C02436
[37] K. Thompson, N. Zeitlin, and P. Mullenix, "Self Healing Coating/Film Project".
[38] N. Salowitz, S. Misra, M. I. Haider, M. Povolo, and P. Rohatgi, "Investigation into the performance of NiTi shape memory alloy wire reinforced Sn-Bi self-healing metal matrix composite," *Materials*, vol. 15, no. 9, p. 2970, May 2022, doi:10.3390/MA15092970
[39] S. Kim, H. Jeon, J. M. Koo, D. X. Oh, and J. Park, "Practical applications of self-healing polymers beyond mechanical and electrical recovery," *Adv Sci*, vol. 11, no. 16, Apr. 2024, doi:10.1002/ADVS.202302463

[40] R. Das, C. Melchior, and K. M. Karumbaiah, "Self-healing composites for aerospace applications," *Advanced Composite Materials for Aerospace Engineering*, edited by Sohel Rana and Raul Fangueiro, pp. 333–364, Jan. 2016, doi:10.1016/B978-0-08-100037-3.00011-0

[41] W. C. K. Tan, J. C. Kiew, K. Y. Siow, Z. R. Sim, H. S. Poh, and M. D. Taufiq, "Self healing of epoxy composite for aircraft's structural applications," *Solid State Phenom*, vol. 136, pp. 39–44, 2008, doi:10.4028/WWW.SCIENTIFIC.NET/SSP.136.39

[42] L. Guadagno et al., "Cure behavior and mechanical properties of structural self-healing epoxy resins," *J Polym Sci B Polym Phys*, vol. 48, no. 23, pp. 2413–2423, Dec. 2010, doi:10.1002/POLB.22139

[43] X. F. Wu et al., "Electrospinning core-shell nanofibers for interfacial toughening and self-healing of carbon-fiber/epoxy composites," *J Appl Polym Sci*, vol. 129, no. 3, pp. 1383–1393, Aug. 2013, doi:10.1002/APP.38838

[44] J. Araujo-Morera, M. H. Santana, R. Verdejo, and M. A. López-Manchado, "Giving a second opportunity to tire waste: An alternative path for the development of sustainable self-healing styrene–butadiene rubber compounds overcoming the magic triangle of tires," *Polymers*, vol. 11, no. 12, p. 2122, Dec. 2019, doi:10.3390/POLYM11122122

[45] Z. Itam, S. Beddu, D. Mohammad, N. L. M. Kamal, N. A. Razak, and Z. A. A. Hamid, "Simulation of alkali-silica reaction model in a concrete gravity dam at the macroscale and mesoscale," *Mater Today Proc*, vol. 17, pp. 717–726, 2019, doi:10.1016/J.MATPR.2019.06.355

[46] F. A. Gilabert, K. Van Tittelboom, E. Tsangouri, D. Van Hemelrijck, N. De Belie, and W. Van Paepegem, "Determination of strength and debonding energy of a glass-concrete interface for encapsulation-based self-healing concrete," *Cem Concr Compos*, vol. 79, pp. 76–93, May 2017, doi:10.1016/J.CEMCONCOMP.2017.01.011

[47] A. Kanellopoulos, P. Giannaros, D. Palmer, A. Kerr, and A. Al-Tabbaa, "Polymeric microcapsules with switchable mechanical properties for self-healing concrete: Synthesis, characterisation and proof of concept," *Smart Mater Struct*, vol. 26, no. 4, p. 045025, Mar. 2017, doi:10.1088/1361-665X/AA516C

[48] S. Gwon, E. Ahn, and M. Shin, "Water permeability and rapid self-healing of sustainable sulfur composites using superabsorbent polymer and binary cement," *Constr Build Mater*, vol. 265, p. 120306, Dec. 2020, doi:10.1016/J.CONBUILDMAT.2020.120306

[49] S. Shirzad, M. M. Hassan, M. A. Aguirre, L. N. Mohammad, S. Cooper, and I. I. Negulescu, "Rheological properties of asphalt binder modified with recycled asphalt materials and light-activated self-healing polymers," *Constr Build Mater*, vol. 220, pp. 187–195, Sep. 2019, doi:10.1016/J.CONBUILDMAT.2019.05.189

[50] L. Pernigoni, U. Lafont, and A. M. Grande, "Self-healing materials for space applications: Overview of present development and major limitations," *CEAS Space J*, vol. 13, no. 3, pp. 341–352, Jul. 2021, https://link.springer.com/article/10.1007/s12567-021-00365-5

[51] Y. Zhu, K. Cao, M. Chen, and L. Wu, "Synthesis of UV-responsive self-healing microcapsules and their potential application in aerospace coatings," *ACS Applied Materials & Interfaces*, vol. 11, no. 36, pp. 33314–33322, 2019.

3 Nanotechnology in Self-Healing Materials

Ankit Tyagi, Gynanedra Prasad Bagri, Sachin Sirohi, Rohit Sahu, Deepak Singh, and Chenglong Yang

3.1 INTRODUCTION

Self-healing materials, inspired by biological repair processes, autonomously mend damage to extend service life and reduce maintenance costs (Blaiszik et al., 2010). Initially showcased by White et al. (2001) using microcapsule-based polymer healing, these materials play a crucial role in aerospace, automotive, infrastructure, and electronics industries (Yang & Urban, 2013). Nanotechnology plays a crucial role in their development, utilizing nanoparticles and nanoscale engineering to enable fast, effective, and versatile repair (Thakur & Kessler, 2015). Self-healing materials represent a transformative class of engineered substances capable of autonomously repairing damage—such as microcracks, scratches, or delamination—without external intervention. This bio-inspired paradigm, first articulated by White et al. (2001), mirrors the wound-healing mechanisms found in living organisms and has rapidly evolved into multidisciplinary research frontier spanning materials science, chemistry, and mechanical engineering (Wool, R. P., 2008). Self-healing functionality extends the service life of structural components, reduces maintenance downtime, and enhances reliability in critical applications ranging from aerospace and automotive industries to civil infrastructure and consumer electronics.

3.2 HISTORICAL EVOLUTION AND CLASSIFICATION

Research on self-repairing systems began with the incorporation of microcapsules containing healing agents in polymer matrices—a concept introduced by White et al. (2001). When ruptured by crack propagation, these microcapsules release monomeric or oligomeric precursors that polymerize and bridge the damaged region. Intrinsic self-healing mechanisms—relying on reversible covalent bonds or supramolecular interactions—were later developed to achieve multiple healing cycles without embedded capsules (Wool, R. P., 2008). The initial stage of research involved studying extrinsic systems, where microcapsules release healing agents in response to damage (White et al., 2001). Systems that rely on reversible bonds, such as Diels-Alder reactions, allow for repeated healing cycles without the need for external agents (Wool, R. P., 2008).

DOI: 10.1201/9781003639510-3

Self-healing materials are broadly classified into:

- **Extrinsic Systems:** Embedded agents (e.g., nanocapsules, vascular networks) repair damage via monomer release (Brown et al., 2005).
- **Intrinsic Systems:** Reversible chemistries (e.g., hydrogen bonds) facilitate autonomous reformation.

3.3 IMPORTANCE OF NANOTECHNOLOGY

Nanotechnology underpins the next generation of self-healing materials by leveraging nanoparticles, nanostructures, and nanoscale engineering to achieve faster, more efficient, and multifunctional repair processes. Key contributions include:

- **High Surface-to-Volume Ratio:** Nanoparticles provide an extensive interface for healing agent storage and release, ensuring rapid interaction with crack surfaces (Wang et al., 2019). Nanoparticles maximize healing agent storage and release (Hager et al., 2010).
- **Functionalized Surfaces:** Surface modification of nanoparticles (e.g., silane coupling agents, polymer brushes) enhances compatibility with host matrices and can catalyze polymerization reactions at damage sites (Zhang et al., 2020). Modified nanoparticles (e.g., silane-grafted silica) improve matrix compatibility and catalyze reactions (Cho et al., 2009).
- **Stimuli-Responsive Behaviors:** Nanoscale materials can be engineered to respond to external triggers—such as temperature, pH, or light—initiating healing precisely when and where needed (Zhang et al., 2020). Nanomaterials respond to heat, light, or pH, enabling targeted healing.

For example, silica nanoparticles incorporated into epoxy matrices can be functionalized to both promote interfacial adhesion and catalyze healing reactions upon crack formation (Wang et al., 2019). Similarly, hollow nanocarriers—such as layered double hydroxides or metal–organic frameworks—enable multi-cycle healing capabilities through controlled release of healing agents under thermal stimuli (Turner, 2021). For example, graphene oxide nanosheets reinforce polymers while enabling reversible bonding (Zarepour et al., 2023).

3.4 MECHANISMS, APPLICATIONS, AND CHALLENGES

Nanotechnology improves extrinsic systems by incorporating nanoreservoirs and intrinsic systems through dynamic bonds (Billiet et al., 2013). Applications encompass aerospace composites, self-healing concrete, and flexible electronics (Aïssa et al., 2012). Challenges such as scalability and stability are tackled through five subtopics: nanoparticle functions, nanoscale engineering, innovations, comparative analysis, and the integration of artificial intelligence (Diesendruck et al., 2015). Graphene-based nanosheets have been integrated into polymer networks to provide mechanical reinforcement alongside reversible bonding sites, facilitating crack closure under ambient conditions (Zarepour et al., 2023). Nanotechnology enhances both extrinsic and intrinsic mechanisms:

- In extrinsic systems, nanoreservoirs ensure uniform distribution of healing agents and mitigate mechanical weakening typically caused by larger microcapsules (Turner, 2021).
- In intrinsic systems, dynamic covalent bonds and supramolecular motifs at the nanoscale enable rapid reformation of the polymer network after damage, with minimal loss of structural integrity (Zhang et al., 2020). Self-healing nanocomposites are poised to revolutionize multiple sectors.
- **Aerospace:** Damage-tolerant composites for aircraft components that autonomously repair in-flight microcracks.
- **Automotive:** Durable coatings and structural parts capable of sealing scratches and preventing corrosion.
- **Civil Infrastructure:** Concrete and asphalt formulations with bacterial additives and nano-silica that heal cracks in bridges and pavements (De et al., 2018).
- **Electronics:** Flexible, self-healing conductive polymers and soft sensors for wearable devices that maintain functionality after deformation (Khatib et al., 2021).

Despite significant progress, several challenges remain:

- **Scalability:** Synthesizing nanoscale healing agents in large volumes at low cost while ensuring uniform dispersion.
- **Long-term Stability:** Preventing premature healing agent degradation or aggregation over service lifetimes.
- **Mechanical Trade-offs:** Balancing high nanoparticle loadings for healing efficiency against potential embrittlement or viscosity increases (Turner, 2021).

3.5 ROLE OF NANOPARTICLES IN SELF-HEALING

Nanoparticles enable precise healing agent delivery, catalytic activity, and mechanical reinforcement in self-healing materials (Toohey et al., 2007). Table 3.1 summarizes key nanoparticle classes in Self-Healing Materials. Nanoparticles serve as the linchpin of modern self-healing materials, offering unparalleled control over healing agent distribution, activation triggers, and mechanical reinforcement. This section explores three key dimensions.

These diagrams encapsulate how mechanical damage triggers nanoparticle actions—capsule rupture, agent release, catalysis, and crack bridging—culminating in restored structural integrity. Figure 3.1 illustrates these mechanisms, depicting nanocapsule rupture, monomer release, and crack closure in a polymer matrix.

3.5.1 Mechanisms of Nanoparticle-Mediated Healing

- **Encapsulation and Release:** Nanocapsules rupture when cracks exceed their critical stress threshold, releasing healing monomers that diffuse into damage zones and polymerize, bridging microcracks (Turner, 2021). Nanocapsules rupture under stress, releasing monomers that polymerize bridge cracks (Turner, 2021).

TABLE 3.1
Nanoparticle Classes in Self-Healing Materials

Nanoparticle Class	Key Features	Examples	Reference
Silica Nanoparticles	Chemical inertness, high surface area, functionalization	Silane-grafted silica in epoxy	Wang et al. (2019)
MOF Nanocapsules	Porous, tunable pore size, multi-cycle release	ZIF-8 nanocages in coatings	Turner (2021)
Carbon-based Nanomaterials	High strength, dynamic bonding, conductivity	Graphene oxide, CNTs in polyurethanes	Zarepour et al. (2023)
Magnetic Nanoparticles	Magnetic response, localized heating	Fe3O4 in polymers	Crall et al. (2017)

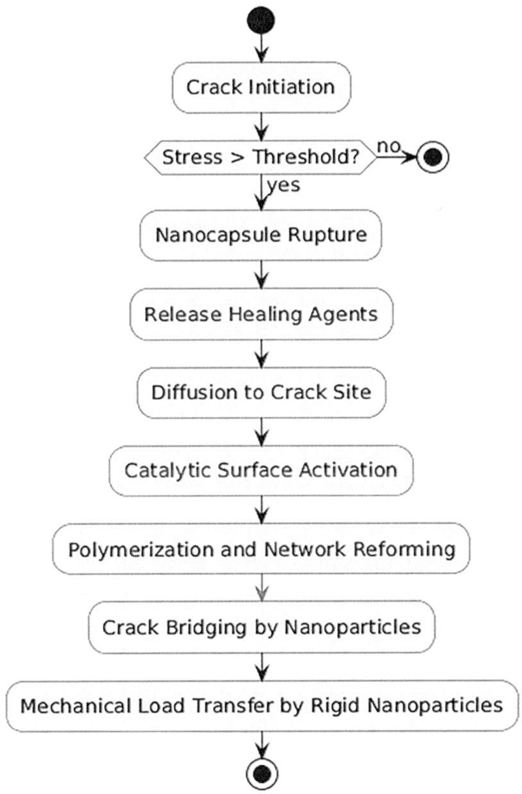

FIGURE 3.1 The mechanisms of nanocapsule rupture, monomer release, and crack closure within the polymer matrix.

TABLE 3.2
Performance of Nanoparticle-Mediated Self-Healing Systems

Material System	Testing Method	Healing Efficiency (%)	Cyclic Healing	Reference
Epoxy + 5 wt% silica	Tensile strength	75	2 cycles	Wang et al. (2019)
ZIF-8 in epoxy coating	Barrier performance	100	3 cycles	Turner (2021)
PU + 0.1 wt% graphene oxide	Tensile, conductivity	85	1 cycle	Zarepour et al. (2023)
Cement + nano-silica, bacteria	Crack closure	65	Multiple	De Belie and Snoeck (2018)
Fe3O4-polymer composite	Tensile strength	80	3 cycles	Crall et al. (2017)

- **Catalytic Activity:** Functional groups grafted onto nanoparticle surfaces initiate polymerization of healing agents upon contact, accelerating repair kinetics (Zhang et al., 2020). Functionalized surfaces (e.g., silane groups) accelerate polymerization (Zhang et al., 2020).
- **Crack-bridging:** Rigid nanoparticles span across fissures, bearing mechanical loads and preventing crack propagation while healing reactions proceed (Wang et al., 2019). Rigid nanoparticles (e.g., graphene) bear loads, preventing crack growth (Wang et al., 2019).

Table 3.2 compares representative nanoparticle-mediated self-healing systems, highlighting their material composition, testing conditions, performance metrics, and associated references.

3.6 NANOSCALE ENGINEERING FOR ENHANCED MATERIAL PROPERTIES

Nanoscale engineering enhances self-healing through tailored surface chemistries, hierarchical structures, and multifunctional designs (Wang et al., 2020). Nanoscale engineering for enhanced material properties combines sophisticated surface functionalization techniques, hierarchical structuring approaches, and multifunctional nano-architectures into a unified framework that significantly elevates both mechanical performance and autonomous repair capabilities; through strategic chemical grafting of silane coupling agents and plasma treatments, nanoparticle and substrate surfaces acquire tailored interfacial chemistries that not only strengthen load transfer but also catalyze healing reactions when damage occurs; concurrently, hierarchical assembly of electrospun nanofibers, carbon nanotubes, and metallic nanowires into aligned or interwoven networks creates tortuous crack paths, distributes stress across multiple length scales, and facilitates heat or electrical stimulus propagation for in situ healing activation; moreover, the integration of core–shell nanoparticles and layered "brick–mortar" composites yields multifunctional systems capable of sensing micro-damage via embedded nanosensors, releasing healing agents through stimuli-responsive mechanisms, and sustaining structural integrity through reversible covalent or supramolecular bonding in a single material platform. Collectively, these nanoscale strategies operate synergistically: surface-functionalized silica and titania nanoparticles grafted with organosilanes bond covalently to polymer matrices, enhancing cohesion while providing active catalytic sites for rapid polymerization; plasma-treated fiber surfaces improve wettability and enable precise deposition of healing monomers; electrospun ceramic nanofibers embedded within polymer matrices deflect crack propagation by forcing microcracks to navigate complex three-dimensional networks; carbon nanotube and graphene oxide reinforcements not only bridge microcracks mechanically but also serve as conductive pathways for Joule heating or electromagnetic induction, triggering thermally responsive healing reactions; finally, core–shell MOF nanocapsules encapsulate diverse healing chemistries within a protective shell that ruptures under mechanical stress or thermal activation, enabling multiple healing cycles without significant performance decline. Below is a summary table of these nanoscale engineering approaches, followed by a PlantUML flowchart illustrating the integrated sequence of sensing, activation, agent release, and network reformation within multifunctional nano-architectures. Table 3.3 shows Nanoscale Engineering Strategies, materials, and benefits. Figure 3.2 describes nanoscale engineering strategies

TABLE 3.3
Shows Nanoscale Engineering Strategies, Materials, and Benefits

Engineering Strategy	Mechanism	Example System	Key Benefit
Surface Functionalization	Chemical grafting (silane coupling), plasma treatment for enhanced interfacial bonding and catalysis	Silane-grafted silica in epoxy	Improved adhesion and accelerated healing kinetics
Hierarchical Structuring	Electrospun nanofiber networks; CNT and nanowire alignment for crack deflection and stimulus delivery	Aligned CNT-infused polymer composites	Multi-scale load transfer; stimulus-responsive healing
Multifunctional Architectures	Core–shell MOF nanocapsules; layered graphene oxide/polymer laminates	ZIF-8 core–shell capsules in protective coatings	Combined sensing, healing, and reinforcement

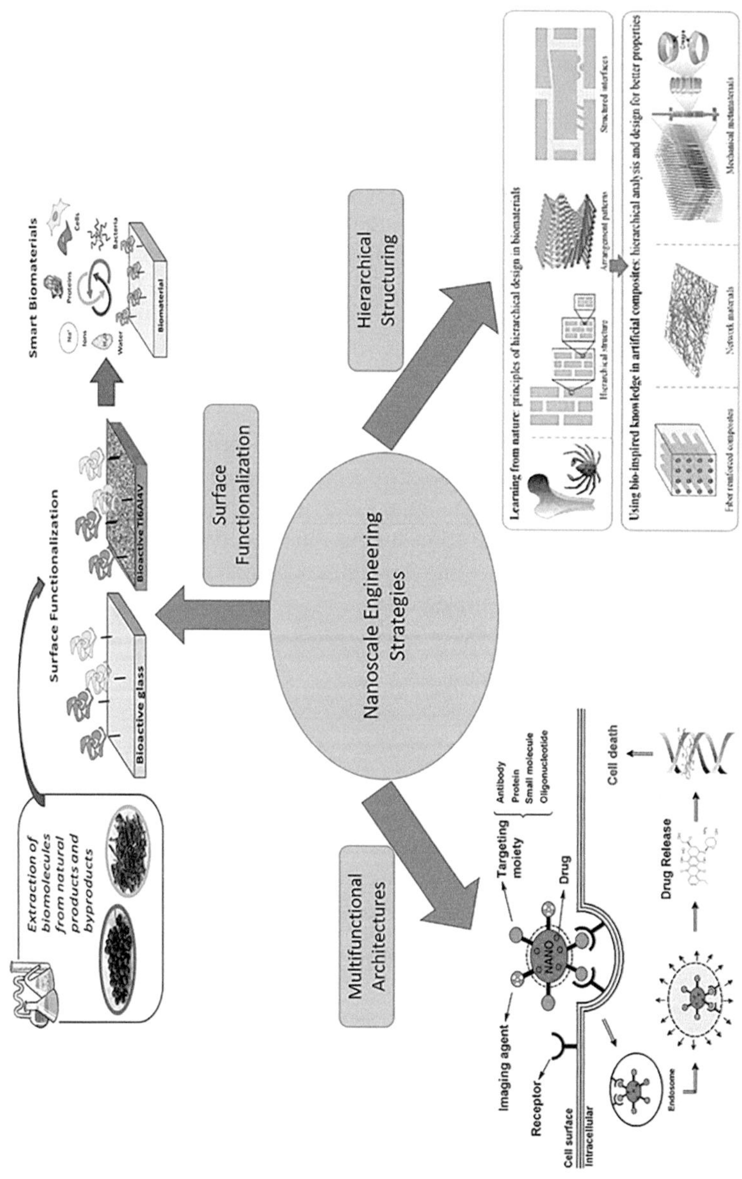

FIGURE 3.2 Nanoscale engineering strategies.

3.6.1 Surface Functionalization

Chemical grafting (e.g., silane coupling) or plasma treatments improve nanoparticle-matrix bonding and catalyze healing (see Section 2.2) (Cho et al., 2009). Silane-grafted silica in epoxy achieves 75% strength recovery (Wang et al., 2019).

3.6.2 Hierarchical Structuring

Aligned nanofibers or carbon nanotubes (CNTs) create tortuous crack paths—complex routes that dissipate stress (Federico et al., 2014). CNT-infused composites use Joule heating (heat from electrical current) for thermal healing (Zarepour et al., 2023).

3.6.3 Multifunctional Architectures

Core–shell MOF nanocapsules release healing agents under stimuli, while layered graphene/polymer laminates combine sensing and repair (Turner, 2021). Figure 3.3 is a flowchart showing sensing, activation, agent release, and network reformation in a nano-architecture.

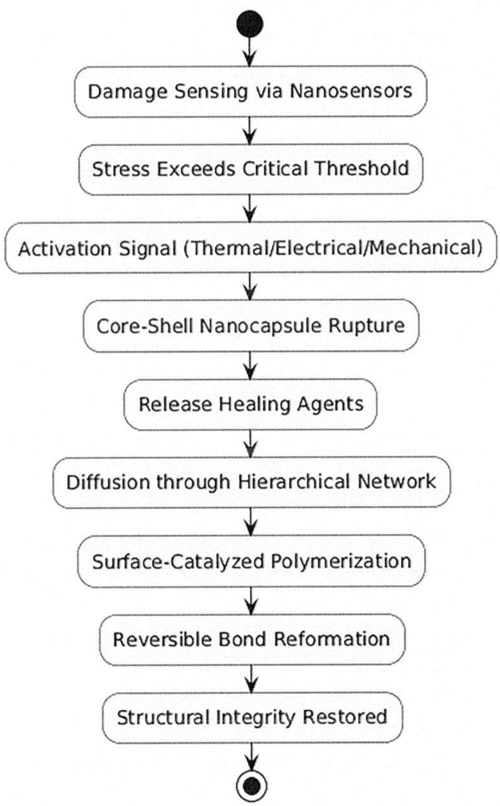

FIGURE 3.3 Flowchart showing sensing, activation, agent release, and network reformation in a nano-architecture.

3.7 INNOVATIONS IN NANO-BASED SELF-HEALING SYSTEMS

3.7.1 Stimuli-Responsive Nanomaterials

Stimuli-responsive nanomaterials leverage environmental triggers—such as pH, temperature, and light—to initiate autonomous healing reactions precisely when and where damage occurs; pH-sensitive polymeric nanogels release encapsulated monomers in acidic or basic environments, while thermoresponsive liposomal nanocarriers undergo phase transitions above critical temperatures to discharge healing agents into microcracks, and photoresponsive nanoparticles containing azobenzene or spiropyran moieties enable light-controlled release and crosslinking of polymer networks, culminating in rapid restoration of mechanical integrity (Zhang et al., 2020). Nanomaterials respond to environmental triggers:

- **pH-Sensitive Nanogels:** Release monomers in acidic/basic conditions, achieving 70% strength recovery (Zhang et al., 2020).
- **Thermoresponsive Liposomes:** Discharge agents above 40°C, healing in 2 hours (Zhang et al., 2020).
- **Photoresponsive Nanoparticles:** Azobenzene-functionalized silica recovers 80% under UV light (Zhang et al., 2020).
- **Magnetic Nanoparticles:** $Fe3O4$ particles induce localized heating for healing (Crall et al., 2017).

3.7.2 Hybrid Healing Strategies

Hybrid systems combine microcapsules and nanoparticles. Epoxy coatings with microcapsules and silica nanoparticles achieve 90% healing over 3 cycles (White et al., 2001; Yang & Urban, 2013). Hybrid healing strategies integrate both microcapsule and nanoparticle concepts to combine the high healing capacity of embedded reservoirs with the mechanical reinforcement and catalytic activity of nanoscale fillers; for instance, microcapsule-laden epoxy coatings supplemented with catalytically active silica nanoparticles achieve simultaneous bulk monomer delivery and surface-accelerated polymerization, resulting in enhanced healing efficiency and multi-cycle durability (White et al., 2001).

3.7.3 Recent Breakthroughs

Recent breakthroughs include cementitious composites incorporating bacteria and nano-silica, wherein ureolytic bacteria precipitate calcite within microcracks and nano-silica provides nucleation sites, achieving up to 65% crack closure in submerged conditions (Zarepour et al., 2023), and graphene-oxide reinforced polymer matrices engineered with reversible Diels–Alder linkages that recover 85% of tensile strength after multiple damage–heal cycles while maintaining electrical conductivity, positioning them as promising candidates for flexible, self-healing electronics (Khatib et al., 2021). Table 3.4 shows the Innovations in Nano-based Self-healing

- **Bio-augmented Cement:** Bacteria and nano-silica precipitate calcite, closing 65% of cracks in 7 days (De Belie & Snoeck, 2018).
- **Reversible Bond Composites:** Graphene oxide with Diels-Alder linkages recovers 85% tensile strength over 5 cycles (Zarepour et al., 2023).

TABLE 3.4
Innovations in Nano-based Self-Healing

Innovation	Trigger/ Mechanism	System	Performance	Reference
pH-Responsive Nanogels	Acid/base	Polymeric nanogels	70% recovery	Zhang et al. (2020)
Bio-augmented Cement	Bacterial precipitation	Cement + nano-silica	65% closure	De Belie and Snoeck (2018)
Biomedical Nanocoatings	Mechanical rupture	Hydrogel + silica	80% scratch repair	Khatib et al. (2021)
Self-Healing Batteries	Mechanical rupture	Graphene electrodes	80% capacity recovery	Yu et al. (2022)
Self-Healing Elastomers	Mechanical rupture	CNT-reinforced elastomers	90% stretchability	Lou et al. (2021)

- **Biomedical Nanocoatings:** Hydrogel coatings with silica nanoparticles repair scratches in implants, maintaining biocompatibility (Khatib et al., 2021).
- **Self-Healing Batteries:** Graphene-based electrodes with nanoreservoirs restore 80% capacity, ideal for electric vehicles (Yu et al., 2022).
- **Self-Healing Elastomers:** CNT-reinforced elastomers recover 90% stretchability for soft robotics (Lou et al., 2021).

3.8 COMPARATIVE ANALYSIS OF SELF-HEALING TECHNOLOGIES

Self-healing technologies vary by scale and material. Macro-scale systems (e.g., microcapsules) heal large fissures but have slower kinetics, while nano-enabled systems offer rapid, multi-cycle repair (Keller et al., 2007). Self-healing technologies span macro-scale approaches—such as microcapsule- or vascular-network-based systems—and nano-enabled strategies leveraging nanoparticles, each with distinct advantages in healing speed, efficiency, and mechanical retention; macro-systems effectively seal larger fissures but often exhibit slower kinetics and diminished toughness after agent deployment, whereas nano-enabled materials distribute healing agents uniformly, catalyze rapid crack closure, and sustain multiple repair cycles with minimal property loss (Aïssa et al., 2012; Billiet et al., 2013). Furthermore, polymeric matrices offer ease of functionalization and flexibility but may sacrifice thermal or chemical robustness, ceramics deliver high-temperature performance at the expense of activation harshness and brittleness, and metals restore ductility through diffusion-driven self-repair under elevated temperatures but incur higher cost and design complexity (Yu et al., 2022; Cho et al., 2009). To assess comparative performance quantitatively, key metrics include healing ratio (recovered strength or stiffness relative to original), toughness recovery, and cyclic healing capability. Table 3.5 summarizes representative systems across material classes, capturing their healing efficiency, toughness recovery, and cycle endurance. Nano-enabled systems excel in cyclic healing, ideal for electronics and coatings, while macro-systems suit structural composites (Pang & Bond, 2005).

TABLE 3.5
Comparative Performance of Self-Healing Systems

System Type	Healing Ratio (%)	Toughness Recovery (%)	Cyclic Healing	Reference
Microcapsule-Based Polymers	60–80	50–70	1	White et al. (2001)
Vascular-Network Ceramics	70–90	60–80	2–3	Turner (2021)
Nano-silica Cement	65	55	Multiple	De Belie and Snoeck (2018)
Graphene-Reinforced Polymers	85	80	5	Zarepour et al. (2023)
CNT-Reinforced Elastomers	90	85	4	Lou et al. (2021)

3.9 AI IN MATERIAL DEVELOPMENT

Artificial intelligence (AI) is transforming the development and deployment of self-healing materials by integrating data-driven nanomaterial discovery, predictive healing kinetics, and real-time system monitoring into cohesive workflows: advanced machine learning models—such as support vector machines, random forests, and deep neural networks—analyze extensive datasets of nanoparticle compositions, surface chemistries, and mechanical test results to rapidly identify optimal formulations and functionalization protocols (Turner, 2021); concurrently, predictive algorithms including feedforward neural networks and Gaussian process regressions model complex, nonlinear interactions among environmental stimuli (temperature, pH), matrix microstructure, and healing agent dynamics, enabling high-fidelity forecasts of healing ratios, toughness recovery, and cycle stability that dramatically reduce experimental iterations; finally, digital twin frameworks link embedded nanoscale sensors (strain gauges, acoustic emission detectors, conductivity monitors) with AI analytics to create live virtual replicas of structural components, facilitating continuous damage detection, threshold-triggered activation of healing mechanisms, and proactive maintenance scheduling, thereby closing the loop between in silico design and in service performance for autonomous lifecycle optimization (De Belie & Snoeck, 2018). AI accelerates self-healing material development through:

- **Material Discovery:** Machine learning (e.g., random forests) predicts nanoparticle–polymer interactions (Turner, 2021; Vasudevan et al., 2021).
- **Predictive Modeling:** Neural networks forecast healing kinetics, optimizing formulations (Turner, 2021; Zhang et al., 2020).
- **Real-Time Monitoring:** Digital twins with nanosensors detect damage and trigger healing, as in aerospace composites (De Belie & Snoeck, 2018; Lee et al., 2019).

For example, AI-designed graphene polymers achieved 30% higher strength and 50% faster healing (Zarepour et al., 2023). AI also optimizes sustainable synthesis, reducing environmental impact (Gupta et al., 2022).

3.10 CONCLUSIONS AND FUTURE DIRECTIONS

Nanotechnology has revolutionized self-healing materials, allowing for accurate healing, improved characteristics, and intelligent functionalities. Obstacles such as reproducible nanoparticle fabrication and ecological consequences persist. Future research should combine artificial intelligence-driven design with real-time monitoring for adaptive systems and prioritize sustainable manufacturing practices. These advancements will encourage the adoption of eco-friendly infrastructure, biomedical devices, and energy storage solutions. Nanotechnology has significantly advanced self-healing materials by enabling precise control over healing agent delivery, catalyzing rapid crack repair, and reinforcing structural integrity through multifunctional nanostructures; however, challenges such as scalable production of uniformly dispersed nanoparticles, long-term stability of healing agents, and optimization of mechanical–healing trade-offs remain critical research areas. Future efforts should focus on integrating AI-driven design with in situ monitoring to develop fully autonomous, adaptive self-healing systems capable of real-time damage assessment and on-demand repair. Moreover, achieving sustainable, cost-effective manufacturing processes and ensuring environmental compatibility of nano-enabled materials will be essential for widespread industrial adoption. Figure 3.4 shows the application of self-healing nanomaterials along with the future trend.

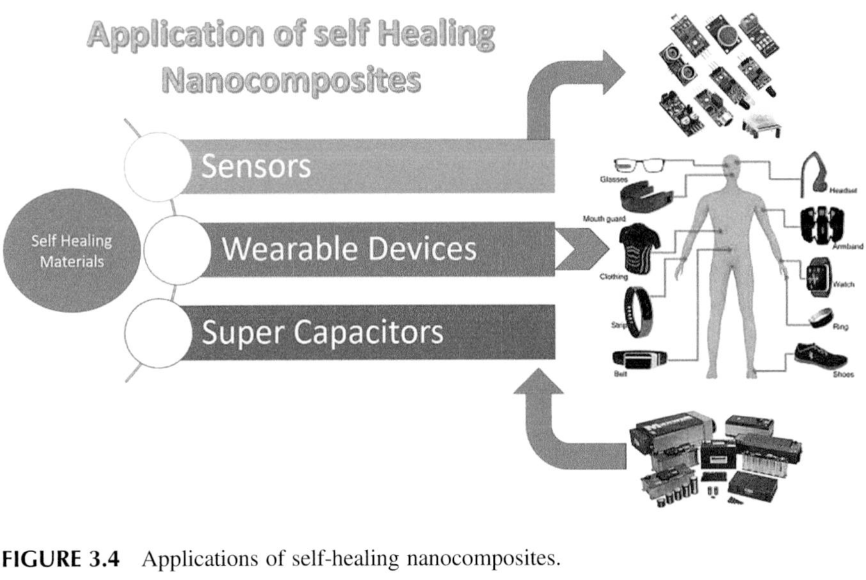

FIGURE 3.4 Applications of self-healing nanocomposites.

REFERENCES

Aïssa, B., Therriault, D., Haddad, E., & Jamroz, W. (2012). Self-healing materials systems: Overview of major approaches and recent developed technologies. *Advances in Materials Science and Engineering*, 2012, 854203. https://doi.org/10.1155/2012/854203

Billiet, S., Hillewaere, X. K. D., Teixeira, R. F. A., & Du Prez, F. E. (2013). Chemistry of crosslinking processes for self-healing polymers. *Macromolecular Rapid Communications*, 34(4), 290–309. https://doi.org/10.1002/marc.201200689

Blaiszik, B. J., Kramer, S. L. B., Olugebefola, S. C., Moore, J. S., Sottos, N. R., & White, S. R. (2010). Self-healing polymers and composites. *Annual Review of Materials Research*, 40, 179–211. https://doi.org/10.1146/annurev-matsci-070909-104532

Brown, E. N., White, S. R., & Sottos, N. R. (2005). Retardation and repair of fatigue cracks in a microcapsule toughened epoxy composite. *Composites Science and Technology*, 65(15–16), 2474–2480. https://doi.org/10.1016/j.compscitech.2005.06.012

Cho, S. H., Andersson, H. M., White, S. R., Sottos, N. R., & Braun, P. V. (2009). Polydimethylsiloxane-based self-healing materials. *Advanced Materials*, 18(8), 997–1000. https://doi.org/10.1002/adma.200501814

Crall, M. D., & Keller, M. W. (2017). Targeted self-healing by magnetically guiding microcapsules. *ACS Applied Materials & Interfaces*, 9(7), 6504–6511. https://doi.org/10.1021/acsami.7b00459

De Belie, N., & Snoeck, D. (2018). Applications of bacteria-based self-healing concrete. *Materials and Structures*, 51(11), 1–11. https://doi.org/10.1617/s11527-017-1130-7

Diesendruck, C. E., Sottos, N. R., Moore, J. S., & White, S. R. (2015). Biomimetic self-healing. *Angewandte Chemie International Edition*, 54(36), 10428–10447. https://doi.org/10.1002/anie.201500484

Federico, B., & Tamer, A. (2014). Self-healing of hierarchical materials. https://doi.org/10.1021/la403497z

Gupta, R., Sharma, S., & Nanda, T. (2022). AI-driven sustainable material design for self-healing applications. *Journal of Cleaner Production*, 340, 130678. https://doi.org/10.1016/j.jclepro.2022.130678

Hager, M. D., Greil, P., Leyens, C., van der Zwaag, S., & Schubert, U. S. (2010). Self-healing materials. *Advanced Materials*, 22(47), 5424–5430. https://doi.org/10.1002/adma.201003036

Keller, M. W., White, S. R., & Sottos, N. R. (2007). A self-healing poly(dimethyl siloxane) elastomer. *Advanced Functional Materials*, 17(14), 2399–2404. https://doi.org/10.1002/adfm.200700086

Khatib, M., Zohar, O., & Haick, H. (2021). Self-healing soft sensors: from material design to implementation. *Advanced Materials*, 33(11), 2004190. https://doi.org/10.1002/adma.202004190

Lee, J., Kim, S., & Park, J. (2019). Digital twins for real-time monitoring of self-healing composites. *Composites Part B: Engineering*, 176, 107321. https://doi.org/10.1016/j.compositesb.2019.107321

Lou, J., Liu, Z., Yang, L., Guo, Y., Lei, D., & You, Z. (2021). A new strategy of discretionarily reconfigurable actuators based on self-healing elastomers for diverse soft robots. *Advanced Functional Materials*, 31(11), 2008328. https://doi.org/10.1002/adfm.202008328

Pang, J. W. C., & Bond, I. P. (2005). A hollow fibre reinforced polymer composite encompassing self-healing and enhanced damage tolerance. *Composites Science and Technology*, 65(11–12), 1791–1799. https://doi.org/10.1016/j.compscitech.2005.03.008

Thakur, V. K., & Kessler, M. R. (2015). Self-healing polymer nanocomposite materials: A review. *Polymer*, 69, 369–383. https://doi.org/10.1016/j.polymer.2015.04.086

Toohey, K. S., Sottos, N. R., Lewis, J. A., Moore, J. S., & White, S. R. (2007). Self-healing materials with microvascular networks. *Nature Materials*, 6(8), 581–585. https://doi.org/10.1038/nmat1934

Turner, A. P. F. (2021). Nanoreservoirs in self-healing coatings. *Progress in Organic Coatings*, 147, 105823. https://doi.org/10.1016/j.porgcoat.2020.105823

Vasudevan, R., Pilania, G., & Balachandran, P. V. (2021). Machine learning for materials design and discovery. *Journal of Applied Physics*, 129(7). https://doi.org/10.1063/5.0043300

Wang, S., & Urban, M. W. (2020). Self-healing polymers. *Nature Reviews Materials*, 5(8), 562–583. https://doi.org/10.1038/s41578-020-0202-4

Wang, X., Zhao, J., & Li, Y. (2019). Nanomaterials and nanostructures in self-healing polymers. *Composites Part B: Engineering*, 175, 107138. https://doi.org/10.1016/j.compositesb.2019.107138

White, S. R., Sottos, N. R., Geubelle, P. H., Moore, J. S., Kessler, M. R., Sriram, S. R., Brown, E. N., & Viswanathan, S. (2001). Autonomic healing of polymer composites. *Nature*, 409(6822), 794–797.

Wool, R. P. (2008). Self-healing materials: a review. *Soft Matter*, 4(3), 400–418. https://doi.org/10.1039/B711716G

Yang, J., & Urban, M. W. (2013). Self-healing polymeric materials. *Chemical Society Reviews*, 42(17), 7446–7467. https://doi.org/10.1039/C3CS60109A

Yu, S., Guo, B., Zeng, T., Qu, H., Yang, J., & Bai, J. (2022). Graphene-based lithium-ion battery anode materials manufactured by mechanochemical ball milling process: A review and perspective. *Composites Part B: Engineering*, 246, 110232. https://doi.org/10.1016/j.compositesb.2022.110232

Zarepour, A., Ahmadi, S., Rabiee, N., Zarrabi, A., & Iravani, S. (2023). Self-healing MXene- and graphene-based composites: properties and applications. *Nano-micro Letters*, 15(1), 100. https://doi.org/10.1007/s40820-023-01074-w

Zhang, Y., Guo, X., Li, H., & Zhao, J. (2020). Stimuli-responsive dynamic covalent hydrogels for self-healing applications. *Polymer Chemistry*, 11, 1–15.

4 Applications of Self-Healing Materials in the Construction Industry

Rina Kamlesh Chokshi, Savita, T. Bernatin, Rohit Sahu, Vijay Chaudhary, Deepak Singh, and Aashay Ingle

4.1 INTRODUCTION

The structural efficiency and long-term reliability of building structures are dependent on the long-lasting nature of the components used in their constructions (Lesovik et al., 2021). During service life, exposure to environmental elements, mechanical loading, and chemical interactions contribute to the processes of microcracking and ultimately material degradation (Thakur et al., 2022), with concrete and asphalt being particularly sensitive. Conventional reparation methods are effective though, they are generally labor-intensive, luxurious, and not resource-efficient (Kumar et al., 2024). Therefore, self-healing materials have been developed as a groundbreaking approach to solve these drawbacks in the construction industry and should be able to autonomously repair damage, thus increasing the inclusive facility life of constructions (Kim et al., 2024). Self-healing materials are developed to recover mechanical performance after sustained damage by various internal or external methods (Qian et al., 2021). The construction industry has made substantial progress in self-healing concrete, where systems introduce healing agents into packages, bacteria spores, or mineral admixtures into the concrete to allow for crack repair and recovery of strength (Sharma et al., 2024). These products have proven that they can decrease permeability, increase durability under repeated loading types of conditions, and delay corrosion in embedded reinforcement to generate more supportable and cost-effective structure solutions (Fernandez et al., 2021). Self-healing technologies not only extend to concrete but also employ self-healing methods in other building materials, certainly, for instance, bituminous pavements, protective coatings (Ammar et al., 2022), fiber-reinforced composites geopolymers, etc. (Feng et al., 2022). The goal of these developments is to reduce maintenance and improve infrastructure resilience, which reduces the lifecycle costs; however, issues still exist regarding material compatibility, healing efficiency, affordability, and the lack of

 DOI: 10.1201/9781003639510-4

standardized test procedures to facilitate widespread applications of these new materials (Alzard et al., 2022).

The present study includes the investigation and optimization of self-healing concrete incorporating bacterial agents, Supplementary Cementitious Materials (SCMs), SMAs, and SAPs, aiming to enhance structural durability, reduce lifecycle maintenance costs, and support sustainable, resilient, and resource-efficient construction practices for long-term infrastructure performance.

4.2 SELF-HEALING MECHANISMS IN CONCRETE

The most used building material in the world, concrete is valued for its consistent adaptability and great compressive strength. However, concrete is a brittle solid and develops cracks for a variety of details, including reduction, mechanical loading, thermal expansion, and environmental conditions (Shaaban et al., 2023). The creation of cracks allows harmful chemicals to enter (water, chlorides, CO_2) into concrete, accelerating the corrosion of steel reinforcement and compromising durability. To address the limitations of concrete and support sustainable infrastructure, self-healing concrete systems have been industrialized, which can autonomously or semi-autonomously repair cracks and restore mechanical function. Figure 4.1 illustrates the self-healing mechanism in concrete. A crack allows water ingress on the right, bacterial spores activated by water precipitate calcium carbonate, sealing the crack. This biological process enhances durability, reduces maintenance, and extends the service life of concrete structures by autonomously repairing micro-damages.

4.2.1 Autogenous Healing

Essentially, this is the continuous drying of untouched cement fragments and carbonation of Ca (OH), whereby this compound responds with CO_2 to create $CaCO_3$, sealing microcracking. As a rule of thumb, autogenous healing occurs primarily

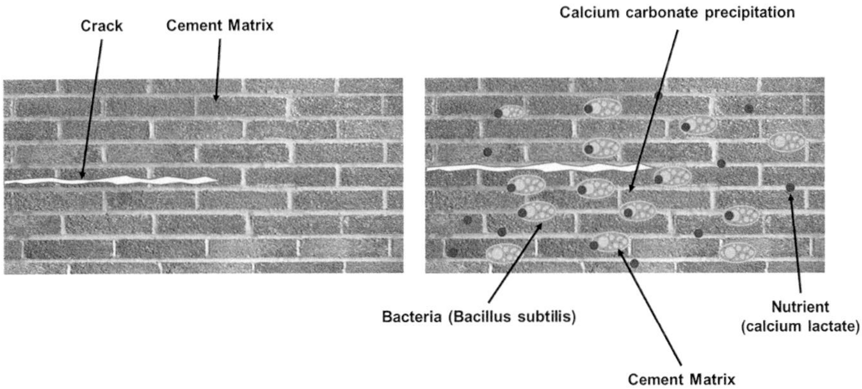

FIGURE 4.1 Self-healing concrete mechanism through bacteria-induced crack sealing.

with cracks ≤ 200 μm under natural environmental conditions (Ghazy et al., 2024). While autogenous healing is inherently passive and cost-free, it is limited in scope by more vigorous or dry conditions and is unable to self-heal when cracks are larger, regardless of the environmental conditions.

4.2.2 Autonomous Healing Mechanisms

Autonomous healing mechanisms are engineered systems intended to extend the self-healing abilities of concrete beyond its natural limits. These methods enable the healing of wider cracks and more diverse environmental conditions. Common approaches included in Figure 4.2 demonstrate the vascular self-healing mechanism in advanced concrete systems. It shows a network of hollow tubes embedded within the concrete matrix, designed to transport healing agents similar to biological circulatory systems. When a crack occurs, the vascular system releases the healing agents into the damaged zone, where they react and fill the void, restoring the material's integrity. This autonomous repair approach significantly enhances the durability, sustainability, and service life of concrete under various mechanical and environmental stresses.

Microbial-Induced Calcite Precipitation (MICP): MICP methods involve incorporating dormant bacterial spores from Velociraptor bacterium, such as *Bacillus subtilis* breeds and nutrients like $C_6H_{10}CaO_6$ in concrete. When cracks form and moisture enters the concrete from the outside, bacteria germinate triggering biochemical processes that lead to the precipitate of $CaCO_3$ (Zhu et al., 2022). This crystalline material seals the crack in the concrete. MICP systems can self-heal cracks as wide as 1 millimeter and improve resistance to water penetration and chemical attack. MICP systems are viewed as very sustainable and environmentally safe, although their performance is only as well as the feasibility of the microbial spores and the circumstances of the local environment.

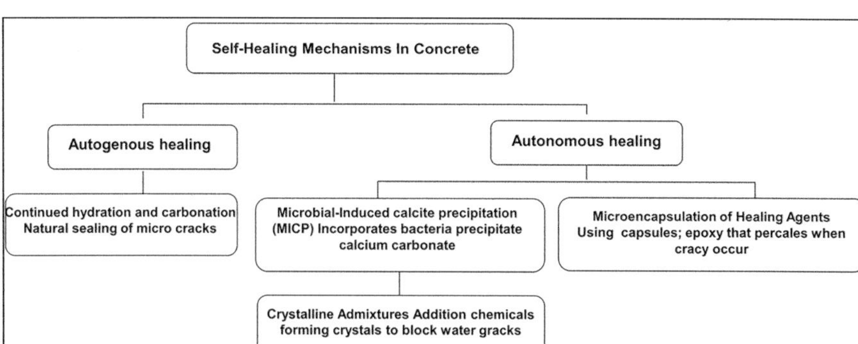

FIGURE 4.2 Self-healing in concrete structure.

Microencapsulation of Healing Agents: The healing agents, which include *epoxy resin, polyurethane, or sodium silicate*, are microencapsulated or wrapped in hollow fibers. The microcapsules are dispersed throughout the tangible environment. When a crack appears in the concrete, the pills break, releasing the healing mediators into the crack. The healing agents epoxy and polyurethane react with moisture in the environment (sodium silicate) to fill the crack and heal it (Hammad et al., 2023). Microencapsulation allows targeted healing, as it can put the healing agents where they are needed, and it can restore both mechanical strength and impermeability. However, production costs and the longevity of the capsules are challenges to their use in practical and large-scale applications.

Crystalline Admixtures: When water permeates through the crack, the activated chemicals will react with the Ca $(OH)_2$ in the pore water, producing insoluble crystals that grow into the crack, and block further water passage. This approach is straightforward, inexpensive, and most suitable for structures that are constantly exposed to moisture. It is most effective in higher moisture condition; however, its healing capacity is limited for wider cracks.

Future Potential: Self-healing systems are a great contributor to sustainability by prolonging service life, reducing maintenance, minimizing lifecycle cost, and lowering environmental consequences (Xiang et al., 2024). Further improvements to materials science, biotechnology, and encapsulation will increase the recital, consistency, and budget-friendly self-healing systems and allow for wider use of self-healing systems in the building industry.

4.3 BACTERIA-BASED HEALING SYSTEMS IN CONCRETE

Incorporating bacterial agents in cementitious matrices provides a biologically based approach toward autogenous crack healing of concrete infrastructure. The activity depends predominantly on uratolytic bacteria, including *Bacillus subtilis*, *Sporosarcina pasteurii*, and *Bacillus megaterium*, which promote $CaCO_3$ through MICP (Zhang et al., 2021). Once microcracks form, in situ environmental parameters (like moisture and air) reactivate the life cycle of dormant bacterial spores. Given an appropriate calcium source and nutrients (e.g., urea, calcium lactate), these microorganisms catalyze biochemical reactions to produce CO_3^{2-}, which can bond with Ca^{2+} for precipitating $CaCO_3$ to fill and seal the crack voids (Su et al., 2024). Viability and efficiency depend upon the bacterial species' ability to withstand alkaline conditions normally present in concrete (pH $\geq$ 11), and the dormant state. Encapsulation technologies, such as porous lightweight aggregates, silica gel, or hydrogel matrices are key to maintaining spore viability and allowing nutrient delivery to the spore. The crack sealing has been shown to be effective for flowblow widths of 0.8 mm and, while the healing was only partial, it was still observed in

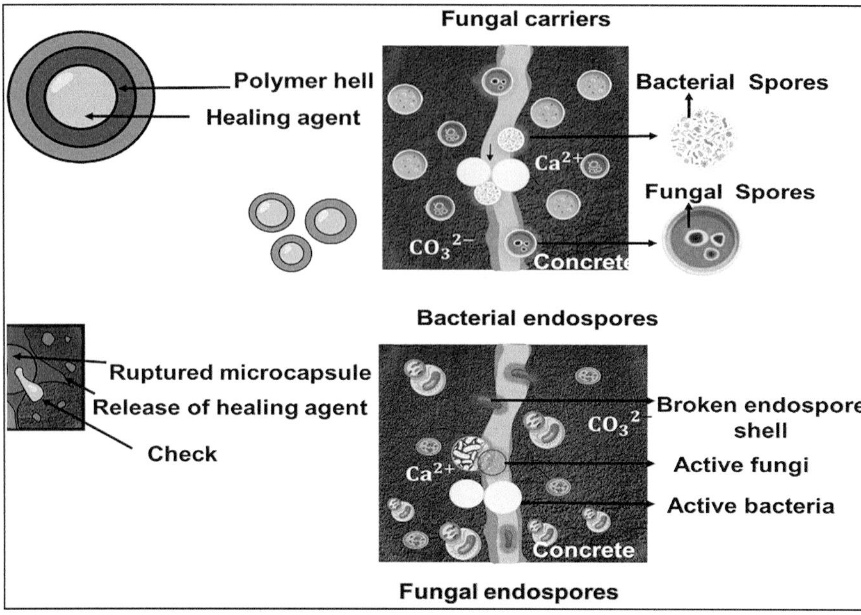

FIGURE 4.3 Bacteria-based self-healing mechanisms for crack repair in concrete.

widths approaching 1.0 mm (Justo-Reinoso et al., 2023). Healing can be influenced by a range of parameters, including bacterial concentration, type and amount of nutrients, aquatic-to-cement ratio, and temperature. The alignment of bacterial systems with SCMs also promotes their inclusion in modern sustainable concrete formulas. Additives, such as fly ash or Ground Granulated Blast Furnace Slag (GGBS) (Vishal et al., 2025) provide pozzolanic activity and improve the pore structure and microenvironment, which promotes microbial activity. When used as a unit, self-healing systems that use bacteria all contribute to longer service life, fewer repairs, and a more durable reinforced concrete product that is subjected to mechanical, thermal, and chemical loads (Elgendy et al., 2025). Figure 4.3 shows different bacteria-based self-healing methods in concrete. The composites displayed consist of an encapsulated healing agent, or bacterial spore with nutrients that are released when the crack has occurred, allowing for the bacteria to become active and produce calcium carbonate ($CaCO_3$) to seal the cracks. In this case, both bacterial and fungal spores can facilitate healing with the spontaneous precipitation of ($CaCO_3$), as water and oxygen become available once mixed. Table 4.1 summarizes the bacterial strains used in self-healing concrete, including information on their survival in alkaline conditions, healing agents, and encapsulation methods. The research reviewed the best microbial systems for self-repairing cracks to increase the durability and sustainability of concrete.

TABLE 4.1
Optimized Comparison of Bacterial Strains for Self-Healing Concrete

Bacterial Strain	Type	Alkaline pH Tolerance	Healing Agent	Max Crack Width Sealed (mm)	Encapsulation Medium	Carbonate Precipitation Efficiency
Sporosarcina pasteurii	Ureolytic	Excellent (≥ pH 11)	Urea + $CaCl_2$	1.0	Silica gel, lightweight AGG	High
Bacillus subtilis	Facultative aerobic	Good (pH 9–11)	Calcium lactate	0.8	Hydrogel microcapsules	Moderate to High
Bacillus megaterium	Non-pathogenic	Moderate	Urea	0.7	Porous ceramic granules	Moderate
Bacillus sphaericus	Anaerobic	Moderate to Good	Urea	0.6	Diatomaceous earth granules	Moderate
Bacillus pseudofirmus	Alkaliphilic	Very High	Urea + Ca acetate	0.75	Pumice aggregate	High
Bacillus cohnii	Thermophilic	Good	Calcium nitrate	0.65	Hollow glass capsules	Moderate
Bacillus circulans	Aerobic	Moderate	Urea	0.5	Expanded perlite	Low to Moderate

4.4 MICROENCAPSULATION TECHNIQUES FOR AUTONOMOUS HEALING

Microencapsulation is a key approach for autonomous self-healing materials to deliver healing agents on demand into microcracks as cracks appear. In cementitious composites, this comes in the form of capsules that contain healing mediators that are buried within the existing method (Yildirim et al., 2022). After crack initiation, the mechanical breakage of the capsules causes the core materials to be spontaneously released, which then fill and seal the crack through polymerization or mineralization processes. Choosing the right capsule shell material is important, as this disturbs the presentation and efficiency of the self-healing process (Parameshwaran et al., 2021). Urea-formaldehyde, melamine-formaldehyde, polyurethane, and silica-based polymers are common shell materials, each with particular mechanical limits and a balance of chemical stability related to specific concrete environments. The shells must endure mixing and curing procedures while being weak enough to fail at stress concentrations due to cracking (Choudhury et al., 2021). The efficacy of the remedial mechanism depends on whether the capsules are spatially distributed and present in the matrix. Consistent distribution is essential to enhance the chances that cracks will intersect with the capsules. Zhu et al. (2024) recommend an optimized dosage of the capsules at 5–10% by weight of the cementitious content in the composite, which significantly enhances the healing effectiveness without compromising the composite's mechanical properties. Trigger instruments are mainly mechanical, relying on induced strain to cause the capsule wall to rupture; however, some systems are triggered by more than just mechanical changes (i.e., adjusting pH, moisture, or temperature in the system). The synergistic effect in dual-capsule systems, in which different capsules contain distinctly separate reactive components that interact and take effect once both capsules have ruptured, provides another sophisticated route to prolong healing ability and recover strength (Xu et al., 2023). Ultimately, microencapsulation creates a passive, scalable self-repair strategy and ties into the goals and vision of sustainable infrastructure-related aspirations, such as long-term durability, reduced maintenance, and an increased service life of concrete behavior.

4.5 SUSTAINABLE USE OF WASTE MATERIALS IN CONCRETE

The use of agricultural and industrial waste by-products in cementitious composites is a growing area of interest that can potentially improve the self-healing recital of tangible while supporting sustainability and the circular economy (Ferroukhi et al., 2023). These waste-derived materials contain a substantial amount of reactive silica and pozzolanic materials, which enhance durability and mechanical performance and the self-healing ability of concrete. Some of the most commonly used materials include GGBS Fly Ash (FA), Sugarcane Bagasse Ash (SCBA), Coffee Husk Ash (CHA), and Rice Husk Ash (RHA) (Chen et al., 2022). These by-products have high pozzolanic reactivity, and it will, cause secondary hydration reactions when they react with calcium hydroxide in the cement matrix. This results in additional calcium silicate hydrate (C–S–H) gel—the main binding phase—which can contribute to self-healing by blocking microcracks and improving pore structure (Nilimaa, 2023).

RHA, which is very fine and has high amorphous silica content, was shown to increase autogenous healing of mortar and concrete samples by allowing continued hydration of anhydrate cement. Fly ash, which is a class F pozzolan, helps in the same manner by forming additional cementitious phases that bridge and close microcracks. CHA, although not as commonplace, provides a mixture of silica, alumina, and carbonaceous constituents, which can support mineral growth in cracks when exposed to moisture. SCBA and GGBS yield both chemical and physical benefits to the research. SCBA is known for its high surface area and porosity, which can aid in internal curing, while GGBS can provide latent hydraulic reactivity, which can initiate strength build-up over time and provide sealing capabilities (Sharma et al., 2022). These materials can improve the material's resistance to aggressive agents like chlorides and sulfates, which have a significant role in improving the service life of concrete structures in aggressive environments. Furthermore, this material also supports the sustainable development of construction practices by diverting large amounts of waste away from landfills and reducing the need for producing Portland cement, as this production generates high energy requirements and CO_2 emissions. The combination of sustainable processes and improved performance provides an environmentally acceptable and technically excellent pathway for the growth of advanced self-healing concrete schemes that rely on using waste products in their chemical composition. This dual-benefit approach reduces harm to the environment and allows the concrete to become more intelligent and durable through healing processes (Shehata et al., 2022). Figure 4.4 illustrates the production of geopolymer concrete using agricultural waste materials, such as rice husk, sugarcane bagasse, and coffee husk. These biomass wastes undergo chemical treatment to

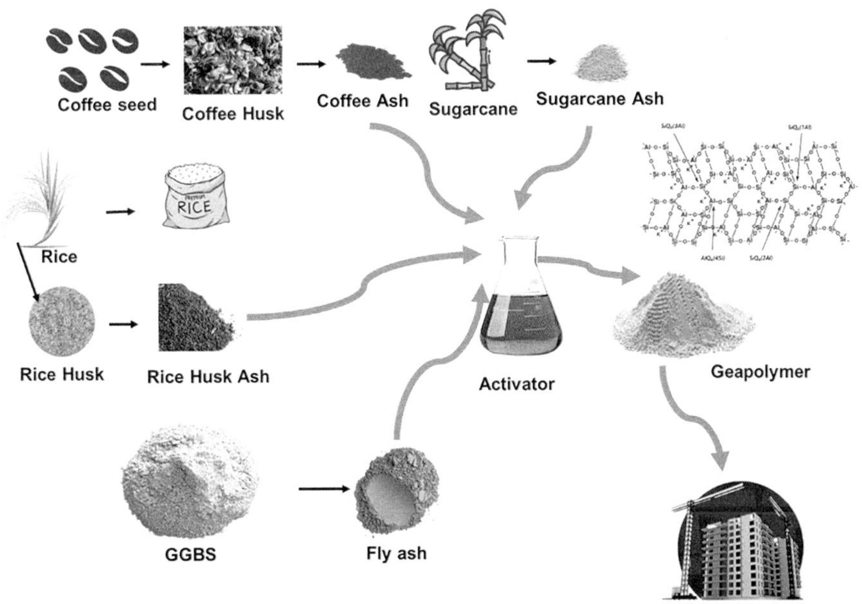

FIGURE 4.4 Geopolymer concrete synthesized from agricultural waste-derived precursors.

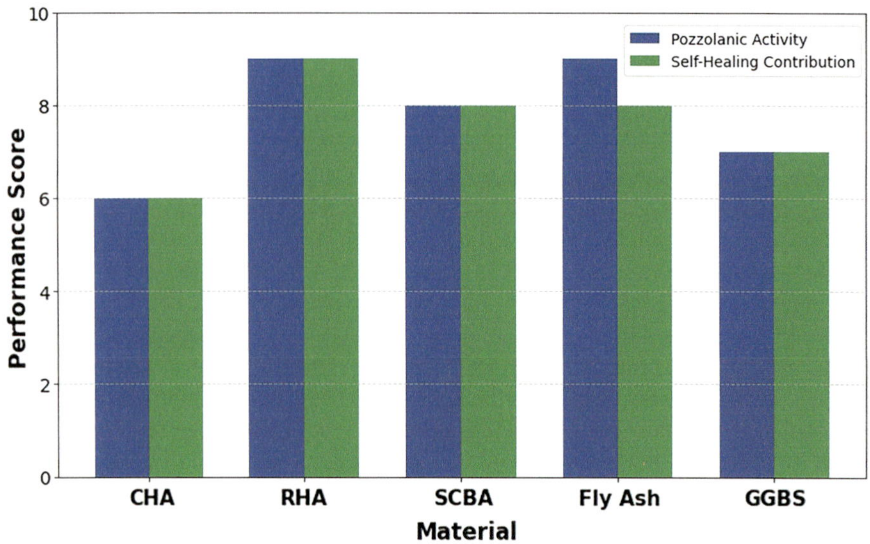

FIGURE 4.5 Comparative performance of waste materials in concrete.

extract reactive silica and alumina, which are used in the polymerization process. The resulting geopolymer binder serves as an eco-friendly alternative to conventional cement, contributing to sustainable construction by reducing carbon emissions and recycling agro-waste into valuable building materials.

Figure 4.5 compares several key waste materials related to their pozzolanic activity and role in self-healing in concrete. These resources provided enhancements in their waste to environmentally friendly concrete and their autonomous self-healing properties. Utilizing waste materials can significantly enhance the performance and durability of concrete while promoting sustainability.

4.6 POZZOLANIC ACTIVITY AND ITS ROLE IN SELF-HEALING CONCRETE

Pozzolanic materials, when used in concrete, significantly upgrade durability and self-healing properties, among other things. Examples of pozzolanic materials are fly ash, RHA, and GGBS (He et al., 2021). Pozzolanic materials contain reactive silica or alumina, which when combined with $Ca\ (OH)_2$ in a cement paste, will react and produce a *C-S-H* gel. C-S-H gel is the primary binding phase in concrete, which will improve concrete density and structure, and strengthen the material (Benameur et al., 2024). The pozzolanic response arises when formless silica from the pozzolanic material interacts with calcium hydroxide formed during the hydration of the cement. This reaction occurs and produces more *C–S–H* gel that will fill microcracks in the carbon dioxide; if this occurs, it demonstrates the concrete's self-healing ability. When microcracks occur, the *C–S–H* is exposed and interacts with the moisture in the environment. This process allows the reformation of new *C–S–H* gel to cover the cracks and sometimes reinstate the original capacity of the concrete itself

(Andriamanalina et al., 2023). The self-healing capability of the substance maximizes the lifespan of the concrete structure by limiting water entry, stopping the propagation of cracks, and improving its resistance to environmental damage. These qualities also enhance the overall sturdiness of concrete, making it more sustainable, especially in infrastructure applications with often severe service conditions. Aside from the motorized and durability assistances of pozzolanic materials, they lessen the environmental impacts of concrete production. Replacing a portion of Portland cement with waste-derived pozzolans reduces concrete's carbon footprint and support more locally/consciously constructed infrastructure. While most pozzolans come from the waste of industrial processes, using these pozzolanic materials tends to be circular by nature and more environmentally responsible than traditional cement construction (Jeon et al., 2022). Incorporating pozzolanic elements into self-healing concrete not only enhances its durability but also contributes significantly to the development of sustainable building solutions for robust infrastructure. Table 4.2 and Figure 4.6 highlight the contribution of industrial and agricultural waste materials to self-healing concrete. These materials, rich in reactive silica and alumina, engage in pozzolanic reactions with calcium hydroxide to form *C–S–H* gel, effectively sealing cracks. Their application not only enhances structural durability but also promotes environmental sustainability by reducing waste and conserving raw materials.

4.7 ADVANCED ADDITIVES FOR ENHANCED INTRINSIC SELF-HEALING MECHANISMS

In traditional self-healing approaches, self-healing in cementitious resources was largely founded on autogenous mechanisms, such as continued hydration or carbonation. However, the use of progressive functional additives can improve the competence and responsiveness of self-repair in concrete structures, which has gained considerable interest (Sarkar et al., 2024). SMAs and SAPs are two recent and promising technologies that accomplish intrinsic healing via physical and chemical mechanisms. Figure 4.7 illustrates a self-healing concrete system incorporating Shape Memory Alloys (SMAs) and Superabsorbent Polymers (SAPs). Upon crack formation, SMA wires contract under thermal activation to close the crack mechanically, while SAPs absorb moisture, swell, and help seal the crack chemically. This dual mechanism enhances the durability of concrete by autonomously addressing damage caused by mechanical stress, temperature fluctuations, and environmental exposure, reducing maintenance costs and extending service life.

Shape Memory Alloys (SMAs) for Active Crack Closure: SMA especially Nickel-Titanium (NiTi) shape memory alloys are unique materials that exhibit thermally induced phase transformations. When embedding SMAs in concrete, they will return to their undeformed shape when heated, generating recovery forces that actively close formed cracks. This not only restores continuity to the structure but can also seal cracks from the environment (Pezzin, 2023). SMAs can be effectively employed in structures that experience thermal cycling, mechanical fatigue, and/or seismic loading. This gives an intelligent option for actively controlling crack development.

TABLE 4.2
Comparative Analysis of Waste-Based Pozzolans in Self-Healing Concrete

Material	Source	Chemical Composition	Pozzolanic Reactivity	Role in Self-Healing	Mechanism of Action	Environmental Impact
Fly Ash	Industrial By-product	SiO_2 (50–60%), Al_2O_3 (20–30%), Fe_2O_3 (5–10%)	High	Reacts with Ca $(OH)_2$ to form C–S–H gel, enhances crack healing.	Alkali-silica reaction (ASR)	Reduces CO_2 emissions, conserves raw materials
RHA	Agricultural Waste	SiO_2 (80–90%), Al_2O_3 (5–7%), Fe_2O_3 (2–4%)	Very High	Forms C–S–H, improving crack sealing due to high silica content.	Pozzolanic reaction with Ca $(OH)_2$	Utilizes agricultural waste, reduces the environmental burden
CHA Materials	Agricultural Waste	SiO_2, Al_2O_3, Fe_2O_3, K_2O, CaO	Moderate to High	Ca $(OH)_2$ to form extra C–S–H gel, promotes crack healing.	C–S–H formation, enhanced crack closure	Recycles agricultural by-products, reduces landfill waste
GGBS	Industrial By-product	SiO_2 (30–40%), CaO (40–50%), Al_2O_3 (10–15%)	Moderate to High	Reacts with Ca $(OH)_2$ to form C–S–H, enhances mechanical strength, and reduces permeability.	Pozzolanic activity with Ca $(OH)_2$	Reduces cement demand, minimizes CO_2 emissions
SCBA	Agricultural Waste	SiO_2 (60–70%), Al_2O_3 (10–12%), Fe_2O_3 (2–5%)	Moderate	Contributes to self-healing by forming C–S–H gel through interaction with Ca $(OH)_2$.	Pozzolanic reaction, crack repair	Supports circular economy, reduces agricultural waste

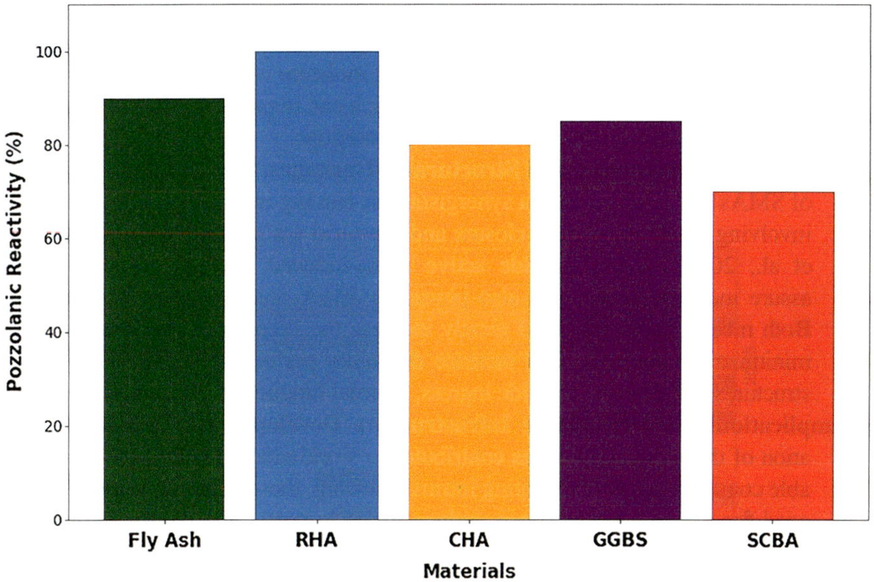

FIGURE 4.6 Crack healing efficiency using pozzolanic waste materials.

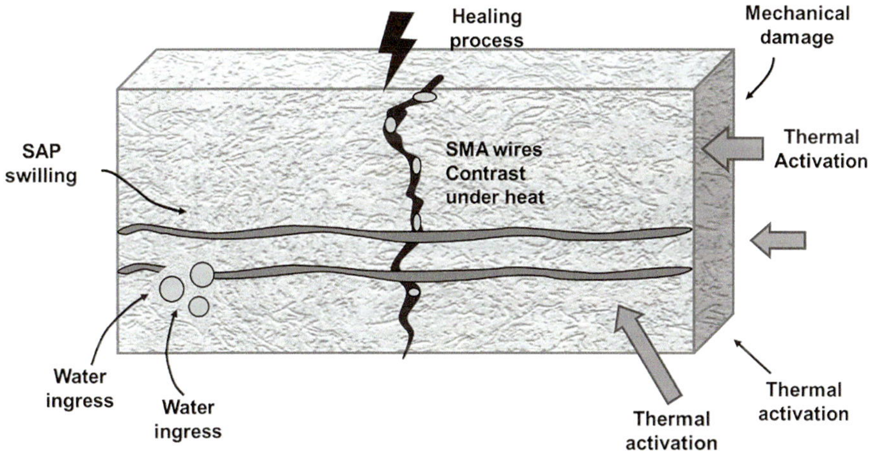

FIGURE 4.7 Self-healing concrete using SMA and SAP dual mechanism.

Superabsorbent Polymers (SAPs) for Moisture-Assisted Healing: SAPs are hydrophilic, cross-linked polymers (also referred to as superabsorbent polymer gels) that can recall large capacities of water compared to their mass. Adding them to concrete allows them to perform two functions: SAPs contribute to internal curing as they announce the engrossed water to assist hydration of the cement particles over a longer duration while also

and lifespan extension were typically characteristics that were showing large increases in anticipated lifespan. This supports the research by quantifying the time-based financial and ecological assistance of self-healing tangible. The evolving benefits of the materials are an integral component in the representation of reduction in expense and improvement in performance over time.

4.9 DISCUSSION

The transformative potential of self-healing concrete in advancing sustainable infrastructure is enormous. Bacterial systems, microencapsulated agents, and vascular networks can greatly facilitate the autonomous recovery of structural integrity through in situ healing of microcracks. Bacteria-based systems can effectively precipitate calcium carbonate within a damaged area, sealing the crack and improving impermeability (Ghazy et al., 2024). At the same time, the use of microencapsulated healing agents and biomimetic networks allows for a continuous and targeted release of healing agents so that maximum functional performance can persist for longer periods of time. Furthermore, by using SCMs, such as GGBS, RHA, SCBA, CHA, and FA, advantages in both ecological sustainability and performance of the material can be achieved (Nilimaa, 2023). These materials contribute to additional C-S-H gels that equal a denser and more durable matrix that improves mechanical stress and chemical ingress resistance due to high pozzolanic activity and reactive silica content (Justo-Reinoso et al., 2023). The capacity to impact self-healing behavior is evident in enhanced closing rates and durability indices. Moreover, the usage of SMAs and SAPs allows for a responsive mechanism with mechanisms including SMAs allowing for crack closure when activated by temperature change and SAPs allowing for internal curing and moisture retention. Together, they work with other self-healing mechanisms and methods, especially in changing environmental conditions. The combined impact of the systems contributes to reduced maintenance frequency and cost and to reduced CO_2 emissions associated with maintaining damage (Jiang et al., 2023). The findings suggest that self-healing concrete, coupled with multifunctional additives and materials derived from waste, could transform the field of concrete design and provide an alternative resilient and resource-efficient approach for next-generation infrastructure projects (Kim et al., 2023; Sarkar et al. 2024).

4.10 CONCLUSION

The integration of self-healing materials in the construction industry represents a transformative step toward sustainable and resilient infrastructure. By incorporating biological agents, encapsulated chemicals, and advanced materials, such as SMAs and SAPs, concrete structures gain the capacity to autonomously address microstructural damage. This intrinsic ability not only enhances mechanical strength and crack resistance but also significantly extends service life while reducing maintenance interventions and environmental impact. The strategic use of agricultural and industrial waste materials, including RHA, fly ash, CHA, SCBA, and GGBS further amplifies pozzolanic reactivity, improves crack-sealing capabilities, and contributes to environmental conservation through circular economy practices. Quantitative assessments confirm notable reductions in lifecycle costs, service downtime, CO_2 emissions, and energy consumption. These developments collectively establish self-repairing concrete, as a crucial development for attaining long-lasting, effective, and sustainable building methods, highlighting its essential function in redefining contemporary civil engineering materials. The high initial costs and the complexity of integrating self-healing systems into large-scale construction projects can pose significant challenges to their widespread adoption. Variability in healing performance due to environmental conditions and material compatibility also presents challenges. Future research should focus on optimizing cost-effectiveness, scaling techniques for field applications, and developing multifunctional materials that enhance both structural and environmental performance across diverse construction settings.

REFERENCES

Alzard, M. H., El-Hassan, H., El-Maaddawy, T., Alsalami, M., Abdulrahman, F., & Hassan, A. A. (2022). A bibliometric analysis of the studies on self-healing concrete published between 1974 and 2021. *Sustainability*, *14*(18), 11646. https://doi.org/10.3390/su141811646

Ammar, M., Haleem, A., Javaid, M., Bahl, S., & Verma, A. S. (2022). Implementing Industry 4.0 technologies in self-healing materials and digitally managing the quality of manufacturing. *Materials Today: Proceedings*, *52*, 2285–2294. https://doi.org/10.1016/j.matpr.2021.09.248

Andriamanalina, T. M., Andrianandrasana, M. D., Raharison, M., & Andrianarison, E. R. (2023). Self-healing concrete by microorganisms in the altered Pozzolan of Madagascar with calcium from crab wastes. *American Journal of Education and Technology*, *2*(1), 1–9. https://doi.org/10.54536/ajet.v2i1.1061

Benameur, S. A., Siad, H., Lachemi, M., & Houmadi, Y. (2024). Development of sustainable self-healing engineered cementitious composites incorporating maximum amounts of natural pozzolan. *Construction and Building Materials*, *456*, 139278. https://doi.org/10.1016/j.conbuildmat.2024.139278

Chen, H., Chow, C. L., & Lau, D. (2022). Developing green and sustainable concrete in integrating with different urban wastes. *Journal of Cleaner Production*, *368*, 133057. https://doi.org/10.1016/j.jclepro.2022.133057

Choudhury, N., Meghwal, M., & Das, K. (2021). Microencapsulation: An overview of concepts, methods, properties, and applications in foods. *Food Frontiers*, *2*(4), 426–442. https://doi.org/10.1002/fft2.94

Elgendy, I. M., Elkaliny, N. E., Saleh, H. M., Darwish, G. O., Almostafa, M. M., Metwally, K., . . . & Mahmoud, Y. A. G. (2025). Bacteria-powered self-healing concrete: Breakthroughs, challenges, and prospects. *Journal of Industrial Microbiology and Biotechnology*, *52*, kuae051. https://doi.org/10.1093/jimb/kuae051

Feng, C., Zong, X., Cui, B., Guo, H., Zhang, W., & Zhu, J. (2022). Application of carrier materials in self-healing cement-based materials based on microbial-induced mineralization. *Crystals*, *12*(6), 797. https://doi.org/10.3390/cryst12060797

Fernandez, C. A., Correa, M., Nguyen, M. T., Rod, K. A., Dai, G. L., Cosimbescu, L., . . . & Glezakou, V. A. (2021). Progress and challenges in self-healing cementitious materials. *Journal of Materials Science*, *56*, 201–230. https://doi.org/10.1007/s10853-020-05164-7

Ferroukhi, M. Y., Mesticou, Z., & Larbi, A. S. (2023). Impact of microencapsulated phase change material on a bio-based building composite, hygrothermal and mechanical behavior. *Construction and Building Materials*, *409*, 133925. https://doi.org/10.1016/j.conbuildmat.2023.133925

Ghazy, A. H., Emara, M. R., Abdellah, A. M., & Attia, M. I. E. S. (2024). Innovative Strategies for Enhanced Concrete Durability: A Comparative Analysis of Autogenous and Autonomous Self-healing Mechanisms. *Iranian Journal of Science and Technology, Transactions of Civil Engineering*, *48*(6), 3927–3953. https://doi.org/10.1007/s40996-024-01426-5

Hammad, N., Elnemr, A., & Shaaban, I. G. (2023). State-of-the-art report: the self-healing capability of alkali-activated slag (AAS) concrete. *Materials*, *16*(12), 4394. https://doi.org/10.3390/ma16124394

He, Z., Shen, A., Wu, H., Wang, W., Wang, L., Yao, C., & Wu, J. (2021). Research progress on recycled clay brick waste as an alternative to cement for sustainable construction materials. *Construction and Building Materials*, *274*, 122113. https://doi.org/10.1016/j.conbuildmat.2020.122113

Jeon, S., Hossain, M. S., Han, S., Choi, P., & Yun, K. K. (2022). Self-healing characteristics of cement concrete containing expansive agent. *Case Studies in Construction Materials*, *17*, e01609. https://doi.org/10.1016/j.cscm.2022.e01609

Jiang, Y., Ng, E. L. L., Han, D. X., Yan, Y., Chan, S. Y., Wang, J., & Chan, B. Q. Y. (2023). Self-healing polymeric materials and composites for additive manufacturing. *Polymers*, *15*(21), 4206. https://doi.org/10.3390/polym15214206

Justo-Reinoso, I., Arena, N., Reeksting, B. J., Gebhard, S., & Paine, K. (2023). Bacteria-based self-healing concrete – A life cycle assessment perspective. *Developments in the Built Environment*, *16*, 100244. https://doi.org/10.1016/j.dibe.2023.100244

Kim, S., Jeon, H., Koo, J. M., Oh, D. X., & Park, J. (2024). Practical applications of self-healing polymers beyond mechanical and electrical recovery. *Advanced science*, *11*(16), 2302463. https://doi.org/10.1002/advs.202302463

Kim, Y. N., Jeong, H., Ryu, S., & Jung, Y. C. (2023). Self-healing polymers and composites for additive manufacturing: materials, properties, and applications. In *Nanotechnology-based additive manufacturing: Product design, properties and applications* (Vol. 1, pp. 219–248). https://doi.org/10.1002/9783527835478.ch8

Kumar, E. K., Patel, S. S., Liao, C. Y., Zhang, L., Hu, S. Y., Xia, S. J., & Li, D. M. (2024). Recent advances of self-healing materials for civil engineering: models and simulations. *Buildings*, *14*(4), 961. https://doi.org/10.3390/buildings14040961

Lesovik, V., Fediuk, R., Amran, M., Vatin, N., & Timokhin, R. (2021). Self-healing construction materials: The geometric approach. *Sustainability*, *13*(16), 9033. https://doi.org/10.3390/su13169033

Nam, J. S., Choi, J. M., Lee, J. W., Han, J., Jeon, I., & Kim, H. D. (2024). Decoding polymeric additive-driven self-healing processes in perovskite solar cells from chemical and physical bonding perspectives. *Advanced Energy Materials*, *14*(12), 2304062. https://doi.org/10.1002/aenm.202304062

Nilimaa, J. (2023). Smart materials and technologies for sustainable concrete Qconstruction. *Developments in the Built Environment*, *15*, 100177. https://doi.org/10.1016/j.dibe.2023.100177

Parameshwaran, R., Naresh, R., Ram, V. V., & Srinivas, P. V. (2021). Microencapsulated bio-based phase change material-micro concrete composite for thermal energy storage. *Journal of Building Engineering*, *39*, 102247. https://doi.org/10.1016/j.jobe.2021.102247

Pezzin, S. H. (2023). Mechanism of extrinsic and intrinsic self-healing in polymer systems. In *Multifunctional epoxy resins: Self-healing, thermally and electrically conductive resins* (pp. 107–138). Singapore: Springer Nature Singapore. https://doi.org/10.1007/978-981-19-6038-3_4

Qian, C., Zheng, T., Zhang, X., & Su, Y. (2021). Application of microbial self-healing concrete: Case research. *Construction and Building Materials*, *290*, 123226. https://doi.org/10.1016/j.conbuildmat.2021.123226

Ravitheja, A., Reddy, T. C. S., & Sashidhar, C. (2022). Improvising the self-healing capabilities of concrete using different pozzolanic materials and crystalline admixtures. *Journal of the Wuhan University of Technology-Materials Science Edition*, *37*(3), 429–439. https://doi.org/10.1007/s11595-022-2549-4

Sarkar, M., Maiti, M., Mandal, S., & Xu, S. (2024). Enhancing concrete resilience and sustainability through fly ash-assisted microbial biomineralization for self-healing: from waste to greening construction materials. *Chemical Engineering Journal*, *481*, 148148. https://doi.org/10.1016/j.cej.2023.148148

Shaaban, S., Hammad, N., Elnemr, A., & Shaaban, I. G. (2023, June). The efficiency of bacteria-based self-healing mechanism in concrete. In *Materials science forum* (Vol. 1089, pp. 135–143). Trans Tech Publications Ltd. https://doi.org/10.4028/p-tc6w54

Sharma, R., Jang, J. G., & Hu, J. W. (2022). Phase-change materials in concrete: Opportunities and challenges for sustainable construction and building materials. *Materials*, *15*(1), 335. https://doi.org/10.3390/ma15010335

Sharma, S. N., Prajapati, R., Jaiswal, A., & Dehalwar, K. (2024, June). A comparative research of the applications and prospects of self-healing concrete/discrete and self-sensing concrete. In *iop conference series: Earth and environmental science* (Vol. 1326, No. 1, p. 012090). IOP Publishing. https://doi.org/10.1088/1755-1315/1326/1/012090

Shehata, N., Mohamed, O. A., Sayed, E. T., Abdelkareem, M. A., & Olabi, A. G. (2022). Geopolymer concrete as green building materials: Recent applications, sustainable development, and circular economy potentials. *Science of the Total Environment*, *836*, 155577. https://doi.org/10.1016/j.scitotenv.2022.155577

Su, Y., Qu, F., Zhang, J., & Zhang, X. (2024). A new bacteria-based self-healing system triggered by sulfate ion for cementitious material. *Journal of Building Engineering*, *86*, 108886. https://doi.org/10.1016/j.jobe.2024.108886

Thakur, A., Kaya, S., & Kumar, A. (2022). Recent innovations in nano container-based self-healing coatings in the construction industry. *Current Nanoscience*, *18*(2), 203–216. https://doi.org/10.2174/1573413717666210216120741

Tomczak, K., Jakubowski, J., & Kotwica, Ł. (2021). Key factors determining the self-healing ability of cement-based composites with mineral additives. *Materials*, *14*(15), 4211.

Vishal, A., Chepuri, A., & Chandana, N. (2025). Assessment of bacteria-based self-healing concrete through experimental investigations—A sustainable approach. *Journal of Materials Science: Materials in Engineering*, *20*(1), 1–16. https://doi.org/10.1186/s40712-025-00215-w

Xiang, J., Qu, L., Fei, X., Qiu, J., & Kong, X. (2024). Depth-dependent self-healing capacity and mechanism of cracked fiber-reinforced concrete by the bacterial community. *Journal of Building Engineering*, *84*, 108485. https://doi.org/10.1016/j.jobe.2024.108485

Xu, D., Huang, Y., Liu, W., Sun, T., & Zhang, M. (2023). Preparation of composite micro-encapsulated phase change material based on phosphogypsum for passive building applications. *Construction and Building Materials*, *378*, 131068. https://doi.org/10.1016/j.conbuildmat.2023.131068

Yildirim, M., Özhan, H. B., & Öz, H. G. (2022). Bacteria-based self-healing of cement mortars loaded at different levels and exposed to high temperatures. *Magazine of Concrete Research*, *75*(13), 674–684. https://doi.org/10.1680/jmacr.22.00238

Zhang, X., Jin, Z., Li, M., & Qian, C. (2021). Effects of carrier on the performance of bacteria-based self-healing concrete. *Construction and Building Materials*, *305*, 124771. https://doi.org/10.1016/j.conbuildmat.2021.124771

Zhu, J., He, J., Zhou, J., Yang, Z., Li, X., Li, Y., & You, Z. (2024). Recent progress in microencapsulation technology and its applications in the petroleum industry. *Journal of Molecular Liquids*, 125162. https://doi.org/10.1016/j.molliq.2024.125162

Zhu, J., Shen, D., Jin, B., & Wu, S. (2022). Theoretical investigation on the formation mechanism of carbonate ion in microbial self-healing concrete: Combined QC calculation and MD simulation. *Construction and Building Materials*, *342*, 128000. https://doi.org/10.1016/j.conbuildmat.2022.128000

5 Applications for Self-Healing Materials in Transportation and Automotive

Sumit Sharma, Kabir Dhingra, Harshit Joshi, Vijay Chaudhary, and Weiguang Fan

5.1 INTRODUCTION

The transportation industry requires smart materials for safer and more sustainable technologies. Self-healing materials is a type of smart materials capable repairing damage autonomously [1]. Their development has been particularly impactful in the automotive sector. They are addressing persistent challenges associated with vehicle advancement.

The self-healing technology becomes essential in engine applications, where materials are subjected to extreme loading conditions, including high temperatures, mechanical stress, and chemical exposure [2]. These advanced materials can be used in critical components such as cylinder liners, piston rings, and valve seats, where microscopic cracks and wear would lead to gradual performance degradation. Self-healing solutions in these applications employ metal matrix composites (MMCs). These MMCs include shape memory alloys or microencapsulated healing agents that activate under specific thermal or mechanical conditions. For example, certain Ni-Ti alloys integrated into engine components [3]. This alloy is also referred to as Nitinol which exhibits shape memory effect. It has a unique property to regain its original shape after deformation when heated, effectively closing fatigue-induced cracks. Sometimes these composites containing microcapsules of healing agent like Sn-Pb alloy [4]. These microcapsules sacrifice when damaged and heated above the melting temperature of the alloy and rescue the damage when cracks propagate through the material.

Self-healing technologies are also making significant impacts across vehicle design. Exterior self-healing coatings capable of repairing scratches and swirl marks not only maintain aesthetic appeal but also preserve corrosion protection. Non-metallic materials like tires with self-healing properties enhance safety by preventing pressure loss from punctures [5]. These applications collectively contribute to extended vehicle lifespans, reduced maintenance requirements, and ultimately the quality of the vehicle.

DOI: 10.1201/9781003639510-5

In aerospace applications, self-healing composites-based systems in structures can detect and repair impact damage, significantly improving aircraft reliability [6]. Marine systems benefit from anti-corrosive coatings that automatically repair scratches, indentation or surface crack in harsh saltwater environments. Infrastructure components like bridges and rail systems are incorporating these materials to stand against fatigue cracks and environmental degradation.

While the potential of self-healing materials is significant, several challenges must be addressed to enable extensive adoption. The current high cost of this type of material presents economic barriers, particularly for mass-market automotive component manufacturing [7]. There are some technical hurdles also related to optimizing healing efficiency under real-world operating conditions, which can ensure consistent performance throughout a product's lifecycle. Furthermore, standardization and testing for these materials are still evolving, requiring close collaboration between researchers, manufacturers, and regulatory bodies [8].

The future of self-healing materials in transportation appears exceptionally promising. Emerging technologies like bio-inspired healing mechanisms and nanotechnology enhanced formulations are pushing the boundaries [9]. The integration of self-healing capabilities with other smart material properties, such as energy harvesting or condition monitoring, may lead to truly multifunctional transportation systems that represent the next leap forward in mobility and automotive industry [10]. This evolution underscores how material science innovations are becoming central to addressing the transportation sector's most pressing challenges while opening new possibilities for vehicle design and optimization.

5.2 SELF-HEALING MATERIALS

Self-healing materials can be broadly categorized based on their composition and healing mechanism. The primary types include those shown in Figure 5.1.

5.2.1 Polymeric Self-Healing Materials

These materials often use embedded microcapsules containing healing agents or possess reversible chemical bonds that allow for reformation after damage [11]. Common applications include automotive coatings and interior components. Polymer-based materials are extensively used in automotive interiors, body panels, coatings, and structural adhesives [12]. With increased use of plastics in interior and exterior body panels offer lower cost and weight, which results in savings for both the manufacturer and the customer (less weight contributes to better fuel economy) emerging as a common practice, the plastic waste generated while the car reaches the end of its life, is considerable [13]. This can be abated by either substituting it with biopolymers, introduce self-healing nature in pre-existing materials or introduce self-healing nature in biopolymers and make them the industry standard. Introduction of self-healing materials will not only improve life of materials in service, increase reliability, and reduce waste but also yield savings for the environment and the economy. Self-healing biopolymers could pave the way for sustainable, eco-friendly development in its truest sense [14].

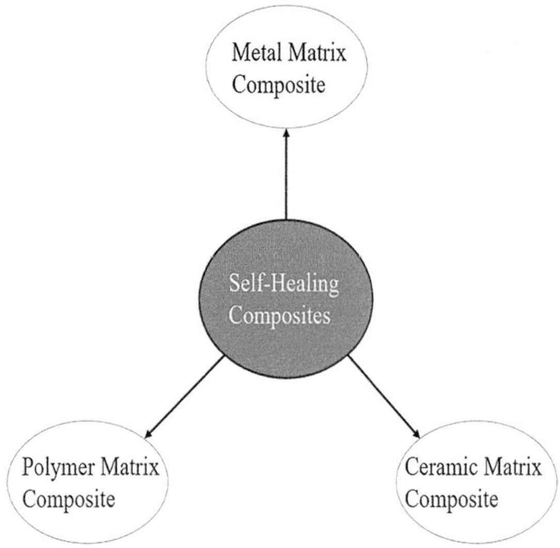

FIGURE 5.1 Broad classification of self-healing composites.

In microvascular systems, a network of tiny channels carries healing agents that flow into cracks when damage occurs, while in microcapsule-based systems, the rupture of capsules releases healing agents directly into the damaged area. Thermally activated composites may use shape memory alloy, polymers, or nanoparticles like graphene or carbon nanotubes that respond to elevated temperatures by softening or flowing to seal cracks, then re-hardening to restore structural strength [15]. In the automotive sector, these self-healing composites are particularly useful in critical areas such as vehicle chassis, body panels, and engine bay components. For instance, carbon fibre-reinforced chassis elements can autonomously repair stress-induced microcracks, while polymer-based panels can self-heal scratches and minor dents, especially when exposed to residual engine heat. By leveraging the vehicle's existing thermal energy, these materials enhance durability, safety, and sustainability by reducing maintenance needs, preventing failure, and extending the overall service life of automotive components.

5.2.2 Metallic Self-Healing Materials

Though still under active research, certain metallic alloys or metal matrix composites exhibit self-healing behaviour through mechanisms like crack tip blunting or diffusion-based healing at elevated temperatures [16]. Metallic self-healing materials are an emerging class of smart materials that demonstrate the ability to autonomously repair damage such as microcracks or fatigue-induced defects, particularly when subjected to elevated temperatures [17]. Although this field is still under active research, significant progress has been made in understanding and engineering alloys and metal matrix composites (MMCs) that can restore structural integrity over time.

One of the primary mechanisms in these materials is diffusion-based healing [18], where atoms migrate to damaged regions when the material is heated, enabling the gradual closure of cracks and the reformation of metallic bonds. This process often occurs at temperatures lower than the material's melting point but high enough to allow atomic mobility.

Another key mechanism is crack tip blunting [19], where the sharp tip of a crack becomes rounded due to local plastic deformation or phase transformations under thermal or mechanical stress. This reduces the stress concentration at the crack tip, slowing its propagation and, in some cases, arresting it entirely. Some metallic systems are also designed to contain embedded healing agents, such as low-melting-point metallic phases, which can flow into cracks upon heating and solidify to restore the load-bearing capability.

Applications of such materials are particularly relevant in high-temperature and high-stress environments, such as automotive engine components, turbine blades, aerospace structures, and nuclear reactors, where regular inspection and repair are challenging [20]. The integration of self-healing mechanisms in metals can significantly extend the operational lifespan of critical components, reduce maintenance costs, and enhance safety. However, challenges remain in terms of healing speed, repeatability, and maintaining mechanical performance after multiple healing cycles. Nonetheless, the advancement of metallic self-healing materials holds great promise for the future of resilient, long-lasting engineering systems.

5.2.3 Ceramic Self-Healing Materials

Used primarily in transportation infrastructure, such as roads and bridges, these materials can heal cracks through the action of embedded bacteria or chemical reactions that precipitate minerals. Cementitious self-healing materials are specially engineered construction materials designed to autonomously repair cracks and micro-damages in concrete structures, significantly enhancing the durability and lifespan of infrastructure [21]. These materials are particularly valuable in transportation-related applications, such as roads, highways, bridges, tunnels, and airport runways, where constant exposure to environmental stresses and heavy loading often leads to cracking and long-term deterioration. The self-healing capabilities in cementitious systems are typically enabled through two main mechanisms: **biological healing** and **chemical autogenous healing**.

In **biological self-healing concrete**, specific strains of dormant bacteria (commonly *Bacillus* species) are embedded in the concrete matrix along with a nutrient source such as calcium lactate. When a crack forms and water enters the concrete, the bacteria become activated, metabolize the nutrients, and precipitate calcium carbonate ($CaCO_3$). This mineral fills the crack and hardens, effectively sealing the gap and restoring the structural integrity of the concrete. This process can continue over multiple cycles, as long as moisture and unreacted nutrients are available.

In chemical self-healing systems, additional components such as crystalline admixtures or encapsulated healing agents are added to the concrete mix. When cracks develop, these agents react with moisture or carbon dioxide in the environment to form solid crystals or gels, which fill the cracks [22]. Another common approach is autogenous healing, where hydrated cement particles in the mix react with infiltrated water over time, leading to the formation of new hydration products that can naturally seal microcracks.

The integration of self-healing mechanisms in cementitious materials offers several benefits, including reduced maintenance and repair costs, extended service life of structures, improved safety, and enhanced resistance to environmental degradation such as corrosion and freeze–thaw cycles [23]. While research and field trials continue to optimize these materials for various climates and load conditions, their adoption in modern civil engineering projects represents a transformative shift towards more sustainable and resilient infrastructure systems.

5.3 INCORPORATING SELF-HEALING PROPERTIES

The integration of self-healing materials into transportation and automotive applications represents a significant leap forward in material science and engineering, promising to enhance the durability, safety, and longevity of vehicles and infrastructure. These materials can be incorporated into vehicles in the following ways.

5.4 INCORPORATION OF SELF-HEALING IN POLYMERS-BASED COMPOSITE COMPONENT

The utilization of self-healing polymers and composites in automotive components, such as body panels, bumpers, and interior parts, can address the common issue of scratches, cracks, and minor damages that vehicles sustain during their operational lifespan [24]. Self-healing polymers, engineered to autonomously repair damage from external stimuli, are at the forefront of sustainable materials research [25]. These advanced materials can extend the lifespan of vehicle components, reduce the frequency of replacements, and minimize waste.

Self-healing polymers have emerged as a promising class of advanced materials for extending the functional lifespan of automotive components. These systems commonly integrate shape-memory fibres and encapsulated healing agents embedded within microcapsules, hollow fibres, or microvascular networks. When mechanical damage occurs, the capsules rupture, releasing the healing agent, typically epoxy resins, dicyclopentadiene (DCPD) with Grubbs' catalyst, polyurethanes, cyanate esters, or polyesters, which then flows into the crack, fills the void, and undergoes polymerization or cross-linking to restore structural integrity. A key implementation is in self-healing automotive coatings, which can be formulated with UV-protective additives and anti-corrosion agents to serve dual functions: repairing surface scratches and mitigating environmental degradation [26, 27]. This not only maintains the aesthetic finish of the vehicle but also safeguards against structural deterioration caused by oxidation, moisture ingress, and photodegradation. In practical applications, such coatings enhance scratch and dent resistance, enabling surfaces to recover from minor abrasions, and improve the service life of adhesives and sealants, where self-healing epoxy systems can restore bond strength after micro-cracking or delamination. Furthermore, recent advances in stimuli-responsive self-healing systems—activated by heat, light, or moisture—have introduced additional performance benefits, including improved thermal stability, chemical resistance, and reduced maintenance demands. By preserving the integrity of both load-bearing and non-structural components, these materials contribute to enhanced vehicle safety, reduced life cycle costs, and greater overall durability [28, 29].

5.5 INCORPORATION OF SELF-HEALING IN METALS MATRIX COMPOSITE

Self-healing systems are particularly well-suited for automotive components that are prone to damage. [30]. These include engine functional metallic components, cylinder block, piston, camshaft, crank shaft, etc. The following objectives can be achieved by incorporating self-healing properties.

a. Reduced weight of the component.
b. Enhanced service life.
c. Higher thermal efficiency

However, its use is not limited to automobiles; it has also been adopted in airplanes, both short- and long-distance trains, and ships for similar benefits. [31]. There following types of self-healing possible in the metallic systems.

5.5.1 Electrochemically Induced Self-Healing

In materials like poly-crystalline metals, they follow up basic diffusion processes like surface diffusion (Ds), grain boundary diffusion (Dgb), vacancy diffusion (Dv), and pipe diffusion (Dp), which are key to how fast atoms move [32–34]. The surface diffusion is faster than the lattice diffusion. This is the reason why models for atomic diffusion and electrochemically induced cell feeling in poly-crystalline materials often combine different diffusion kinetics [1].

Electrochemical cell feeling is a favourable strategy for metals, especially titanium (Ti). For example, it can repair itself by re-oxidising in air if the protective oxide film on the Ti gets damaged.

The cracks on a titanium component can heal when the oxidation reaction products fill them. This means the cracks from the operational stress can autonomously self-heal through re-oxidative reactions in Ti-based materials [18, 35–37]. The challenge of this self-healing approach is that the extent of oxidation depends on the sample temperature.

5.5.2 Micro-encapsulated Healing Agent Reinforced Metal Matrix Composites (MMCs)

It consists of placing small hollow structures like microspheres or micro tubes, containing a low melting metal inside a main metal that melts beta higher temperature. The outer shell of the microcapsules acts as a barrier, but it must also be weak enough to break when a crack reaches it, it can break instead of deflecting the crack

[38]. Researchers attempted to place hollow microfibres filled with low melting healing substances into metal systems. One attempt involved using Indium (a low melting metal) inside carbon tubes embedded in a higher melting solder. After being heated beyond the melting point of Indium, a large downwards-oriented crack was successfully healed [4].

5.5.3 Shape Memory Alloy (SMA) Reinforced Metal Matrix Composites (MMCs)

This method involves inserting wires of shape memory alloy (SMA), such as nitinol (a nickel titanium alloy), into the main metal. These SMA wires have a unique property: when heated about a specific temperature they return to their original shape.

Therefore, when a crack occurs in the metal composite, the stresses surrounding the crack stretch the SMA wires spanning it. When the material is subsequently heated to the SMA's shape transformation temperature, the wires contract to their original dimensions. This contracting process applies a strong compressive force to the metal matrix effectively clamping the crack closed. At the same time, the matrix alloy is designed to melt partially at this healing temperature, allowing the crack to be welded shut [5]. The method of self-healing uses a special type of metal called shape-memory alloys (SMAs), like nickel titanium (NiTi). These metals have the ability to "remember" their original shape. When these alloys are cooled, they are in a softer state known as martensite. If we deform this, martensite can be made to straighten out by a little heat. This heating transforms it back to its original, hard state known as austenite, and it recovers its shape in the process. The researchers demonstrated this thermally engineered self-feeling in action. They made tiny dents in a nickel titanium alloy using different tools. When the alloy was gently heated, the dents nearly disappeared, particularly the circular ones. This proves that the shape memory effect works even on a tiny scale and under different types of pressure comma and the material can heal itself [3].

5.6 HEALING MECHANISMS

The schematic representation in Figure 5.2 illustrates the classification of self-healing mechanisms in materials, broadly divided into intrinsic (non-autonomous) and extrinsic (autonomous) systems.

In intrinsic self-healing mechanisms, the healing capability is embedded within the material's inherent structure and chemical composition, without reliance on external intervention or foreign repair agents. This category is further subdivided into:

Healing Agent-Based Systems – These utilize encapsulated or network-based delivery methods to transport healing substances to the damaged site. Common strategies include *microencapsulation* (where healing agents are enclosed in microcapsules that rupture upon damage), *micro-vascular networks* (mimicking biological circulatory systems to deliver repair agents), and *hollow fibre embedment* (fibres containing liquid agents that are released when fractured).

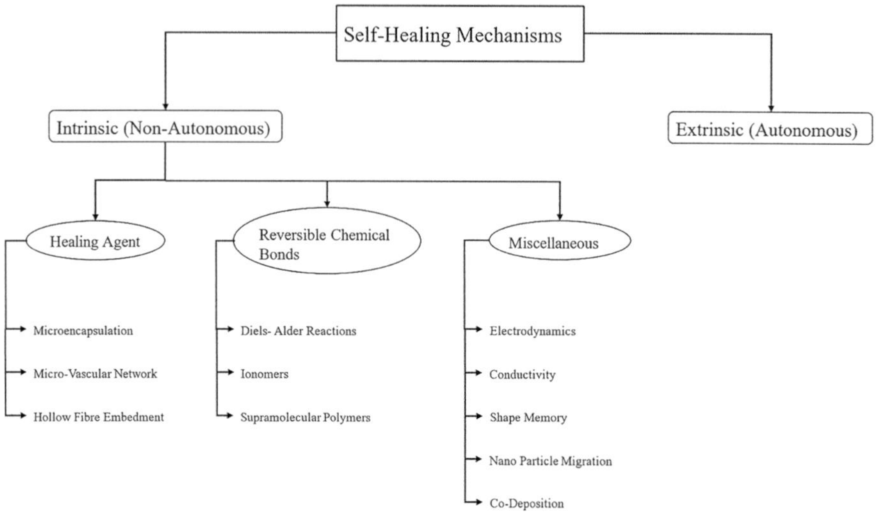

FIGURE 5.2 Healing mechanism.

Reversible Chemical Bond-Based Systems – These mechanisms exploit dynamic covalent or supramolecular interactions that can break and reform under specific stimuli. Examples include *Diels–Alder reactions* (thermally reversible covalent bonding), *ionomer-based systems* (ionic interactions aiding structural recovery), and *supramolecular polymers* (non-covalent interactions such as hydrogen bonding and π–π stacking).

5.7 MISCELLANEOUS MECHANISMS

This group encompasses a variety of physical and chemical recovery processes, such as *electrodynamics* (restoration of electrical pathways), *conductivity recovery*, *shape memory effects* (thermally or stress-activated reformation), *nanoparticle migration* (movement of nanoparticles to damaged sites to enhance bonding), and *co-deposition* (layer formation for surface repair).

Intrinsic self-healing relies on the inherent chemical and physical properties of the polymer matrix, enabling it to repair damage without the need for additional healing agents. This mechanism is primarily driven by reversible bonding interactions or thermally activated molecular mobility within the material. One widely studied approach is based on Diels–Alder (DA) and retro-Diels–Alder (rDA) chemistry [39], where furan and maleimide functional groups form reversible covalent bonds. Upon heating, the rDA reaction breaks these bonds, allowing polymer chains to reorient and diffuse into the damaged zone. Cooling then triggers the forward DA reaction, reforming the covalent network and restoring mechanical properties. This process can be repeated multiple times without significant loss of performance [22].

Other examples of intrinsic healing systems include:

a) **Disulfide Bond Exchange** – Polymers containing dynamic disulfide linkages can heal through reversible bond reshuffling, triggered by heat or UV light. This mechanism is common in elastomers and coatings that require repeated damage recovery.
b) **Hydrogen Bonding Networks** – Strong and reversible hydrogen bonds, such as in ureidopyrimidinone (UPy)-functionalized polyurethanes, enable rapid healing at ambient or slightly elevated temperatures.
c) **Ionic Interactions in Ionomers** – Ionomers like Surlyn utilize ionic cluster reorganization to close microcracks upon heating, often used in golf ball covers and potentially in automotive impact-resistant panels.
d) **Metal–Ligand Coordination** – Coordination bonds between polymer ligands and metal ions (e.g. terpyridine–Zn^{2+}) provide a dynamic network that can self-repair when heated or mechanically stimulated.
e) **Dynamic Imine or Acylhydrazone Bonds** – These reversible covalent bonds allow self-repair under mild acidic or basic conditions, useful for adhesives and sealants in automotive assembly.
f) **Supramolecular Polymers** – Networks held together by non-covalent interactions, such as π–π stacking or host–guest chemistry, can heal through simple molecular rearrangement, often without external heat.

In automotive applications, intrinsic self-healing systems can be incorporated into dashboards, steering wheel coatings, interior trim, sealants, and flexible electronics, where repeated small-scale damage must be repaired without downtime or component replacement. These systems often provide multiple healing cycles, higher durability under thermal cycling, and reduced maintenance costs compared to extrinsic capsule-based approaches.

5.8 EXTRINSIC HEALING

In contrast, extrinsic self-healing mechanisms are autonomous systems that operate independently of the material's inherent chemistry. These typically involve embedded external repair systems or pre-programmed microstructures that detect and address damage without manual intervention.

This classification framework serves as a comprehensive reference for selecting and developing appropriate self-healing strategies for advanced engineering applications, considering both chemical and physical recovery pathways.

Extrinsic self-healing systems rely on embedded reservoirs of healing agents that are released upon damage, enabling repair through chemical reactions or physical filling of cracks. Unlike intrinsic systems, the healing functionality is not inherent to the polymer backbone but is supplied by an external material stored within the structure. This approach is particularly effective for one-time or limited-cycle repair of large-scale damage where rapid restoration of structural integrity is critical.

The most common extrinsic healing strategies include:

a) **Microcapsule-Based Systems** – Healing agents (e.g. epoxy resins, dicyclopentadiene (DCPD), polyurethanes, cyanoacrylates) are encapsulated within spherical shells made of urea–formaldehyde, polyurethane, or melamine–formaldehyde. When cracks propagate, they rupture the capsules, releasing the liquid monomer into the damaged zone where it polymerizes—often initiated by a catalyst embedded in the matrix.
   Example: DCPD with Grubbs' catalyst in epoxy composites for repairing automotive body panels and load-bearing components.
b) **Hollow Fibre Systems** – Healing agents are stored in glass, polymeric, or carbon hollow fibres embedded within the composite. When fractured, the fibres act like pipelines, delivering the healing resin directly into the crack.
   Example: Hollow glass fibres filled with epoxy for repairing damage in automotive carbon-fibre–reinforced bumpers.
c) **Microvascular Networks** – Inspired by biological circulatory systems, interconnected channels are integrated into the material to deliver healing agents from a reservoir. These networks enable multiple healing cycles and can be refilled after each repair event.
   Example: 3D-printed vascular networks in structural frames of electric vehicles for long-term damage management.
d) **Dual-Capsule or Multi-agent Systems** – Two or more types of capsules store separate reactants (e.g. monomer and hardener). Upon rupture, the agents mix and cure within the crack. This method allows for rapid polymerization and improved healing strength.
   Example: Epoxy + amine curing agent dual-capsule systems in adhesive joints of automotive chassis assemblies.
e) **Core–Shell Nanoparticles** – Nanocapsules containing corrosion inhibitors or lubricants can be dispersed into protective coatings, releasing their contents when surface microcracks form.
   Example: Nanocapsule-loaded paints for automotive exteriors to prevent rust formation in scratched regions.
f) **Solid-State Healing Fillers** – Low-melting-point alloys or thermoplastic particles embedded in the composite melt upon heating and flow into cracks, resolidifying to restore structural continuity.
   Example: Thermoplastic microbeads in crash-resistant car frames to improve post-impact recovery.

In automotive applications, extrinsic systems are widely used in structural composites, exterior coatings, adhesives, sealants, and corrosion-protective paints. Their primary advantage is the ability to address large-scale or visible damage quickly, but they are often limited to single-use repairs unless designed with refillable vascular systems. Combining extrinsic healing with intrinsic mechanisms can create hybrid systems that offer both immediate repair capability and long-term durability.

5.9 RECOVERY ANALYSIS AND MECHANISM OF SELF-HEALING COMPOSITES

The interaction between SMA fibres and the surrounding matrix—particularly the bonding quality and efficiency of load transfer—is crucial for the functionality of SMA-reinforced SHMMCs. Factors such as the influence of pre-strain in SMA fibres on load transfer, crack propagation rate, crack closure rate, and the extent of shape recovery must be thoroughly evaluated. Manuel [39] introduced an analytical model aimed at estimating the minimum volume fraction of SMA wires necessary to apply sufficient clamping force at the crack interface during a thermal healing cycle. The effectiveness of the healing process in SMA-reinforced SHMMCs is strongly influenced by several parameters, including the proportion of SMA content, the compressive yield strength of the matrix, and the flow stress of the SMA in its parent phase.

5.10 HEALING ASSESSMENT OF SHAPE MEMORY ALLOY REINFORCED SELF-HEALING METAL MATRIX COMPOSITES

The healing efficiency of a composite is assessed by comparing the mechanical properties of the virgin composite with the post-heal mechanical properties of the same composite, using the following relationship

$$\text{Healing efficency} = \frac{\text{Healed measurement}}{\text{Original measurement}} \times 100\ [40]$$

The formula shown in the image is used to calculate the healing efficiency of self-healing materials, particularly in experimental studies involving Self-Healing Metal Matrix Composites (SHMMCs), polymers, or SMA-integrated systems. If the system restores the original shape and size after deformation the healing efficiency is 100% [40].

For an example, analysis of a self-healing composite beam that includes shape memory material (SMM) reinforcement undergoes bending. When the beam is acted on by an external force, it gets deformed. Normally, an external force would be required to again to bring it back to its original shape [41, 42]. Using shape memory materials, that restoring force can come from the material itself. Thus, a self-healing material can be made using shape memory as the healing mechanism [43]. The idea is to figure out how much of this shape memory material is needed to bring the beam back to its original form and how much restoring force is generated by the shape memory reinforcement as equal amounts of energy needs to be provided in the form of heat.

A basic beam bending is explored, and observation is made as to how the internal recovery from SMM matches the external bending force. We'll also talk about stress–strain hysteresis, i.e. the energy loss that happens during this self-healing process and how it affects the design of such a beam with SMM reinforcement.

Post the bending of the beam, shape memory material inside should push back enough to make the beam straight again. The objective is to calculate:

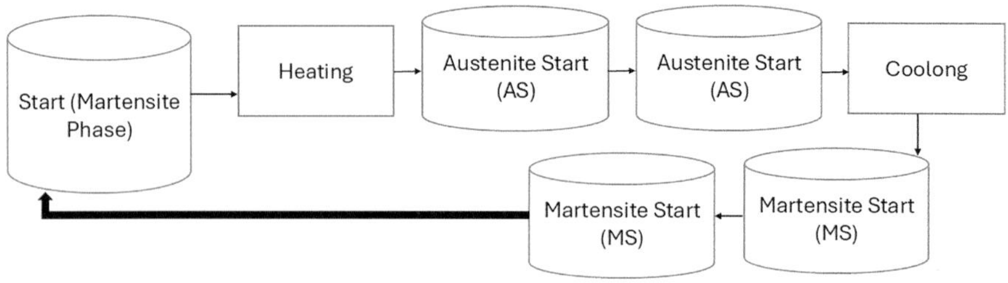

FIGURE 5.3 Stress–strain SMM.

1. How much bending moment the beam experiences.
2. How much restoring force the SMM can provide.
3. How much of the SMM is required for the restoration.

Hysteresis is the phenomenon due to which, the material doesn't return all the energy provided to it. In other words, the stress–strain curve when loading the material isn't the same as when it recovers [44]. That gap, the resulting loop between loading and unloading is energy lost as heat. Figure 5.3 shows Stress–Strain SMM.

Due to hysteresis, we don't assume 100% efficiency and a parameter of recovery efficiency must be included to account for this energy loss [45]. If the stress–strain hysteresis wasn't considered, we'd have underestimated how much SMM we need and the beam might not have fully healed post the deformation [46].

5.11 THE MECHANICS OF SELF-HEALING IN METAL MATRIX COMPOSITES

Self-healing composites especially metal matrix can reinforce SMA fibres like Nitinol ranging from short to long continuous fibres to recover the dimensional integrity after failure. These fibres function by closing cracks, recovering the band or reducing crack growth into the damaged region.

To design such mechanism, it is essential to develop mathematical models that accurately determine the precise SMA fibre volume fraction required to induce yielding in the metal matrix while avoiding damage to the SMA during the healing cycle. Such models must also define the pre-strain levels for the SMA fibres and their optimal placement and orientation within the matrix to maximize healing efficiency.

Important factors that require quantification include:

Quality of interfacial bonding between SMA fibres and the matrix alloy.
Load transfer mechanisms from matrix to fibres, also the state of loading iso-stress/iso-strain.
Crack propagation and shape recovery rate in the composite during the healing process.

Analytical approach estimates the minimum volume fraction of SMA fibre needed to generate sufficient clamping force across a crack interface during healing cycle.

The model includes SMA transformation temperatures range along with the high-temperature compressive yield strength of the matrix. Further it is typically described using the Johnson–Cook constitutional equation, which characterizes the temperature-dependent compressive flow stress of the matrix alloy. This relationship is expressed in Equation (5.1):

$$\sigma_{flow} = [A + B\epsilon^{n}]\,[1 + Cln\dot{\epsilon}]\left[1 - \left(\frac{T - T_{room}}{T_{melt} - T_{room}}\right)^{m}\right] \tag{5.1}$$

Here, flow stress of matrix alloy (σ_{flow}), Equivalent plastic strain (ϵ), Normalized plastic strain rate ($\dot{\epsilon}$), Room temperature (T_{room}), Melting temperature of matrix alloy (T_{melt}), Current temperature of the matrix alloy during deformation or healing (T) [47].

A, B, C, n, and m are analytically derived constants obtained through experimental measurements. These constants ensure the model accurately reflects the material's mechanical response under given loading and environmental conditions.

In Equation (5.1), the flow stress of the alloy is influenced by three primary factors, work hardening, strain rate sensitivity, and thermal softening, each represented by one of the three bracketed terms. When the objective is to isolate and analyse only the temperature-dependent effects on stress, Equation (5.1) can be reduced to a simplified form, presented as Equation (5.2).

$$\sigma_{flow} = A\left[1 - \left(\frac{T - T_{room}}{T_{melt} - T_{room}}\right)^{m}\right] \tag{5.2}$$

The flow stress of the matrix alloy approaches zero as it reaches its melting temperature (T_{melt}). However, for alloys containing a eutectic phase, the relevant softening point occurs earlier—at the eutectic temperature (T_{eut}) [40]. At this temperature, partial liquefaction of the eutectic phase initiates, significantly reducing the alloy's load-bearing capacity during the healing process. To account for this behaviour, the Johnson–Cook model is adapted by replacing T_{melt} with T_{eut}, yielding the modified form shown in Equation (5.3), which accurately represents the thermal softening at the eutectic healing temperature [40].

$$\sigma_{flow} = A\left[1 - \left(\frac{T - T_{room}}{T_{eut} - T_{room}}\right)^{m}\right] \tag{5.3}$$

During the healing phase, shape memory alloy (SMA) wires exert a recovery force that draws the cracked matrix back together at elevated temperatures. At equilibrium, the restoring force generated by the SMA wires balances the opposing force within the matrix. This force balance can be expressed by normalizing the respective stresses to their cross-sectional areas for both the matrix and the SMA wires, as represented in Equation (5.4).

$$F_{SMA} = F_{matrix} \Rightarrow \sigma_{SMA}A_{SMA} = \sigma_{matrix}A_{matrix} \tag{5.4}$$

Dividing Equation (5.4) by the total cross-sectional area gives us a new relationship, presented in Equation (5.5), where V_f is the volume fraction of the SMA wires:

$$\sigma_{\text{SMA}} V_f = \sigma_{\text{matrix}} \left(1 - V_f\right) \tag{5.5}$$

In the healing stage, the stress within the SMA wires is governed by their recovery stress (σ_R), which represents the stress generated as the wires revert to their original shape. Conversely, the stress within the matrix is determined by its compressive yield stress at elevated temperatures (σ_{MCYS}).

For effective healing, the recovery stress of the SMA wires must be sufficient to cause the matrix to yield locally, [40] ensuring crack closure without inducing undesirable plastic deformation in the SMA elements. Based on this requirement, Equation (5.5) can be reformulated into the modified form shown in Equation (5.6).

$$\sigma_R = \sigma_{\text{MCYS}} \left[\frac{Vf}{1 - Vf}\right] \tag{5.6}$$

Equation (5.6) shows how the volume fraction of SMA wires is indirectly related to the recovery stress of the wires and the compressive yield stress of the matrix. Since the flow stress of an alloy is the same as its compressive yield strength (i.e. $\sigma_{flow} = \sigma_{MCYS}$), Equation (5.3) can be rewritten as:

$$\sigma_{\text{MCYS}} = A\left[1 - \left(\frac{T - T_{room}}{T_{eut} - T_{room}}\right)^m\right] \tag{5.7}$$

The analytical model for determining the optimal volume fraction of SMA wires is obtained by substituting the expression for σMCYS from Equation (5.7) into Equation (5.6), resulting in Equation (5.8). This formulation enables the calculation of the SMA wire volume fraction required to generate a sufficient clamping force at the crack interface throughout a healing cycle, ensuring effective crack closure and material restoration.

$$\sigma_R = A\left[1 - \left(\frac{T - T_{room}}{T_{melt} - T_{room}}\right)^m\right]\left[\frac{V_f}{1 - V_f}\right] \tag{5.8}$$

Manuel [39] demonstrated that, for a Sn-based proof of concept composite, a 1 vol% fraction of Ti-49.4 at% Ni SMA wires was sufficient to enable self-healing. In contrast, a Mg-based composite required approximately 30 vol% of the same SMA wires due to the higher strength of the magnesium matrix, which resisted the recovery forces, necessitating a greater SMA content.

Fisher [33] applied the same model to Al-Cu-Si and Al-Si alloys. Predictions indicated that an Al-4.1Cu-2Si (at%) composite would require $\geq$ 55 vol% Ni-Ti SMA wires, exceeding the matrix volume—a surprisingly high ratio. For Al-3 at% Si, the model estimated 25 vol%, yet experimental results showed healing with $<$ 2.5 vol% SMA wires.

This significant mismatch between model predictions and experimental outcomes highlights limitations in the Manuel model, particularly its assumptions of equal strain and absence of matrix plasticity in self-healing metal-matrix composites (SHMMCs). Additionally, improved toughness in synthesized SHMMCs could further reduce the SMA content required for effective healing [40]. Consequently, there is a need for a refined analytical model that accounts for these factors to accurately determine the minimum SMA volume fractions necessary for optimal self-healing performance.

5.12 DEVELOPED MODEL FOR AUTONOMOUS SELF-HEALING

A conceptual model has been developed for autonomous self-healing in automotive systems, enabling onboard repair of damaged components without manual intervention as shown in Figure 5.4. The system integrates microprocessor-based control units, embedded sensors, and actuators, working in coordination to detect damage, evaluate its severity, and initiate the healing process. Heat for repair can be sourced from multiple inputs, including engine heat or electrical resistive heating. The operational flowchart illustrates a shape memory alloy (SMA)-based self-healing process, in which the system first assesses the recoverability of the damaged region. If repair is feasible,

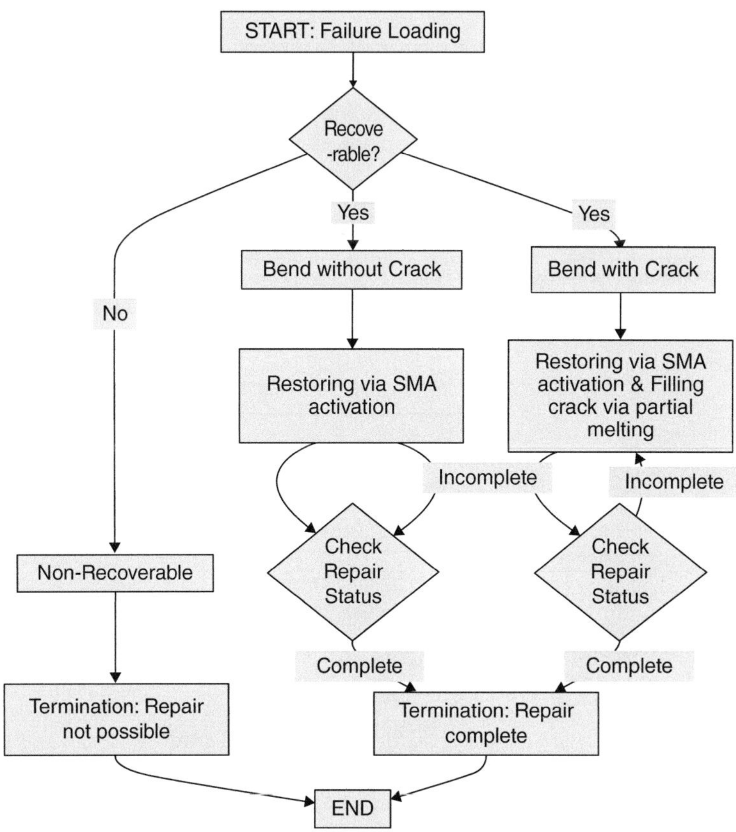

FIGURE 5.4 Recovery analysis loop.

the SMA elements are activated—either by direct heating or partial melting of the polymer matrix—allowing the material to recover its original geometry and restore structural integrity.

The process supports repeated checks and loops until repair is either complete or deemed impossible, promoting enhanced material longevity and safety in engineering applications.

The process begins with a failure event and assesses whether the damage is recoverable. If not, the process ends by declaring the repair impossible. If recoverable, the system distinguishes between two scenarios—deformation without cracking and deformation with cracking. For cases with no cracks, the material is restored using SMA activation, leveraging its ability to return to its original shape upon heating. For cases involving cracks, SMA activation is combined with partial melting to fill and heal the crack. After either process, the system checks the repair status. If incomplete, the repair loop is repeated; if complete, the system terminates the process, confirming that the repair was successful. This method reflects a robust approach to enabling autonomous or semi-autonomous repair mechanisms in smart materials, particularly self-healing metal matrix composites.

5.13 ARTIFICIAL INTELLIGENCE (AI) DRIVEN SELF-HEALING

Self-healing becomes autonomous when it is integrated with AI. The developed model is presented in Figure 5.5. The process begins with the initialization phase, where the material is subjected to controlled deformation to simulate real-world

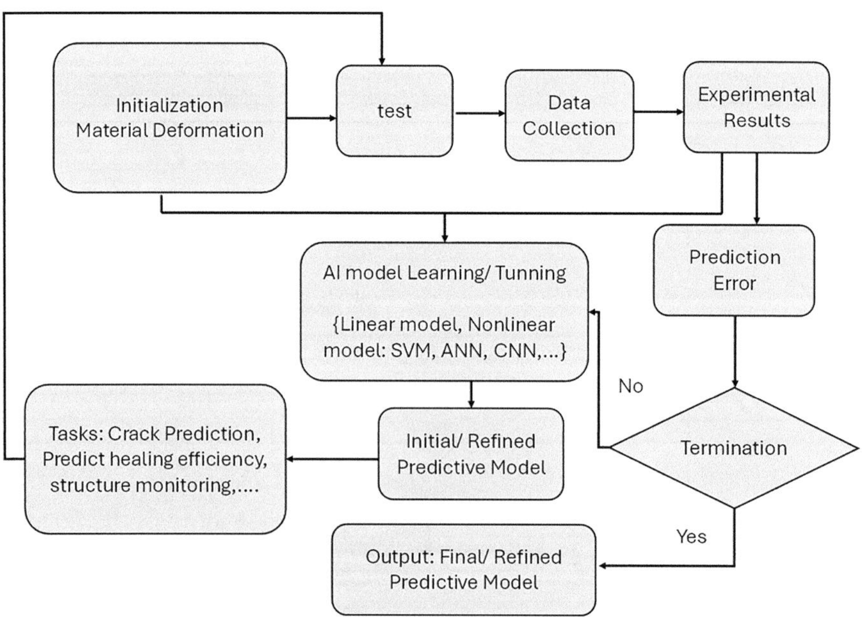

FIGURE 5.5 AI model for self-healing.

stress or damage conditions. This step is crucial for generating scenarios that reflect actual material behaviour. Following this, a testing phase is conducted, which can be experimental or simulation-based, to monitor the material's response under load. During this phase, relevant data such as strain, stress, temperature, crack progression, or acoustic emissions are collected for further analysis.

The collected data is processed to obtain experimental results, which serve as the foundational input for training artificial intelligence (AI) models. The AI model learning or tuning stage involves using both linear and non-linear models depending on the complexity of the data. Common algorithms used include Support Vector Machines (SVM), Artificial Neural Networks (ANN), and Convolutional Neural Networks (CNN), especially for image-based datasets. This phase is iterative, requiring hyperparameter tuning and validation to ensure optimal model performance. The output of the model is compared against actual experimental results to calculate the prediction error.

If the prediction error is within an acceptable range, the process moves to the termination phase, indicating that the model is sufficiently accurate. Otherwise, the model is refined further through retraining or by incorporating additional data. The result is an initial or refined predictive model, which evolves over successive iterations. Once finalized, the model can perform specific tasks such as crack prediction, estimating self-healing efficiency, and monitoring structural integrity. Ultimately, this AI-based predictive framework provides a powerful, data-driven method for evaluating and predicting material behaviour, with applications in fields such as aerospace, automotive engineering, infrastructure, and biomedical devices where reliability and material performance are of critical importance.

5.14 HEAT-DRIVEN CRACK REPAIR IN ENGINE COMPONENTS VIA HIGH-SPEED OPERATION

The concept of self-healing in engine components such as cracked pistons or cylinders is an idea in materials science [48]. It aims to take advantage of the large amounts of heat present in running engines and repair small cracks without having to disassemble the engine in the process. Piston and cylinder running conditions create extremely high thermal and mechanical loads. These loads come from the pressure of combustion, the rapid axial movement of the piston, dust or contaminants present in the fuel-air mixture, and constant friction. These types of loads are all contributing factors to surface or subsurface cracking in piston and cylinder. Between the very high temperatures and high loading conditions; there is a high frequency of crack occurrence in critical regions of the piston such as the crown, ring grooves, and inner walls of the cylinder. These cracks typically do not cause catastrophic engine failure if they are small, but when they exceed the tolerance of the engine the cracks can increase in size and engine damage occurs. Cracks and similar damage to the engine typically require the complete disassembly of the engine and replacement of damaged components. Self-healing proposes to make use of these characteristics and simply repair the damage during normal operation of the engine using the naturally produced heat of the engine.

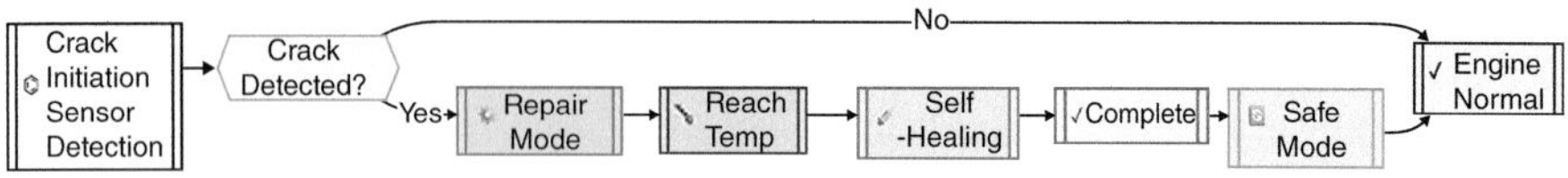

FIGURE 5.6 Crack repair process.

When an engine operates at high RPM, the combustion action increases at a faster rate, thereby also increasing the heat resulting from combustion much more quickly. Some of the heat and energy from combustion goes to heating up the piston and cylinder walls. Additionally, the friction generated when the piston rings meet the cylinder liner creates further heat, which is more pronounced at high RPM. At low-speed or idle, there is not enough heat to enable healing to take effect. However, at higher speeds or load there is sufficient heat and friction such that it raises the temperature of the engine components. This additional heat and friction will increase the tendency for self-healing to take effect. Pistons and cylinders are conventionally made from non-ferrous alloys, such as aluminium-silicon and aluminium-magnesium. When these materials reach high temperature, they include action such as diffusion, creep, sintering, and grain boundary movements; this sometimes facilitates the opportunity to automatically close a small crack. Also, if the components were designed with a specific built-in self-healing agent, such as a low melting alloy (Sn-Bi, Zn-Al), hollow capsules filled with healing fluids or reactive particles, heat from high RPM can activate them. These materials melt and flow into the crack via capillarity or pressure. Once the engine cools sufficiently, the self-healing agents solidify and seal the crack. Figure 5.6 shows the crank repair process.

In shape memory alloys, temperature may induce a martensite to austenite phase transformation. It aids the material in going back to its initial shape and healing cracks. High temperatures can also induce strengthening in certain materials through precipitation of particles or sintering of oxide layers. In coated cylinder liners, it may prevent the extension of cracks. For surface cracks, temperature can also facilitate the rebuilding of protective films or nano-crystalline films that are designed for self-healing [49].

This technique suits very small or surface cracks. It doesn't repair deep or big cracks. Success with self-healing relies on a lot of factors: how long the heat remains, how big and where the crack is located, how hot the engine components become, and what materials are involved. Engine operation needs to be tightly controlled. Too much heat for too long can lead to further damage rather than healing [50, 51].

5.15 APPLICATIONS IN AUTOMOTIVE INDUSTRY

Self-healing materials have found various applications in the automotive sector, including:

Coatings: Self-healing paints and clear coats can automatically repair minor scratches and abrasions, preserve the vehicle's aesthetic appeal and protect underlying materials.

Self-healing coatings are commonly classified based on their mode of self-repair, into physical (e.g. diffusion and randomization) and chemical (e.g. chemical reactions, polymerizations.) processes

Interior Components: Dashboard surfaces, seating materials, and control panels benefit from self-healing polymers that maintain appearance and functionality over time.

Structural Components: Advanced composites used in car bodies and frames can self-heal, improving crash resistance and prolonging structural integrity under cyclic loading.

Battery and Electronic Systems: Research is underway to develop self-healing electrodes and circuit pathways that enhance the durability and safety of electric vehicle batteries.

5.16 APPLICATIONS IN TRANSPORTATION INFRASTRUCTURE

Beyond automobiles, self-healing materials play a vital role in other modes of transportation:

Railways: Self-healing concrete and composites are used in railway ties and structural components to reduce maintenance needs and improve safety.

Aerospace: Lightweight self-healing composites are used in aircraft fuselages and wings to enhance performance and reliability during long flights.

Maritime: Ship hulls and offshore platforms benefit from corrosion-resistant self-healing coatings that extend operational lifespans.

Road Infrastructure: Self-healing asphalt and concrete can automatically repair micro-cracks, reduce potholes, and enhance road safety.

Some of the real-world applications of smart materials in the automotive industry are (Figure 5.7):

- **Nissan Scratch Shield:** Nissan together with Advanced Soft materials Inc. and University of Tokyo developed an original Scratch Shield paint. It was earlier used for automobiles such as Nissan X-Trail, Murano and 370Z. Later, it was also used to make iPhone cases. The Scratch Shield paint repairs smaller scratches in approximately an hour but as the localization

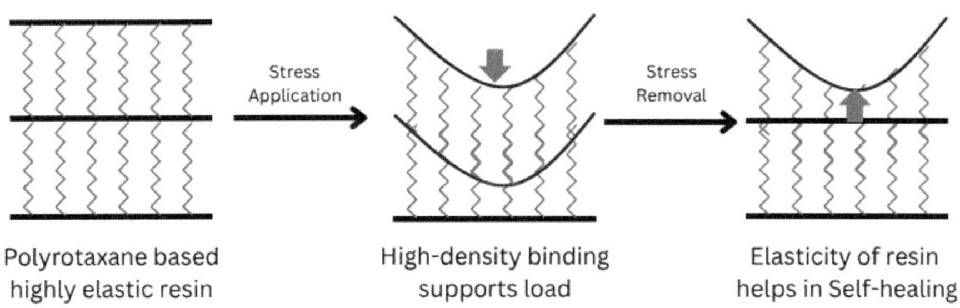

FIGURE 5.7 Nissan Scratch shield.

increases, the duration of healing will also increase. The Scratch Shield paint utilizes extrinsic self-healing with polytrioxane used as a healing agent in the spaces of the elastomeric matrix.

Lamborghini Smart Material System (LSMS): Lamborghini's LSMS was developed for its hybrid super sports cars: Lamborghini Sián Roadster and Lamborghini Sián FKP 37. The LSMS is placed on the rear side just below the rear wing. It utilizes thermos-induced shape memory alloy springs to operate four flaps on the outlet on the engine cover. As the under-hood temperature increases, the SMA springs expand resulting in the opening of flaps, which leads to better cooling performance. When the under-hood temperature decreases, the SMA springs contract resulting in the closing of flaps, which leads to better aerodynamic performance. It is advantageous as no electric, hydraulic, or mechanical actuation is needed which leads to weight saving. Lamborghini, in collaboration with MIT, has been exploring self-healing paint technology for future supercars. The concept, exemplified by the Lamborghini Terzo Millennio, involves using electricity-storing carbon nanotubes embedded within composite panels.

5.17 CHALLENGES AND FUTURE OUTLOOK

Despite their promising advantages, self-healing materials face several challenges:

Cost: The incorporation of self-healing mechanisms can be expensive, limiting widespread adoption in cost-sensitive applications.
Scalability: Producing self-healing materials on an industrial scale with consistent performance is a technical hurdle.
Durability of Healing Mechanism: Ensuring that the healing function remains effective over the lifecycle of the material is crucial.
Environmental Stability: The performance of self-healing materials under varying environmental conditions needs further study [52].

Future research is focussed on developing multi-functional self-healing systems that not only repair themselves but also respond to stimuli such as temperature, light, or stress. With advancements in nanotechnology, smart polymers, and additive manufacturing, the scope for self-healing materials in transportation and automotive applications is expected to expand significantly [53].

5.18 CONCLUSION

Self-healing materials represent a significant advancement in the pursuit of intelligent and sustainable design in the transportation and automotive sectors. Their ability to autonomously repair damage can lead to safer, more durable, and cost-effective systems. While challenges remain in terms of cost, scalability, and long-term performance, ongoing research and technological progress continue to push the boundaries of what is possible. The integration of these materials into everyday applications promises a future where maintenance is minimized, safety is enhanced, and material longevity is maximized.

REFERENCES

[1] Z. G. Wei, R. Sandstroröm, and S. Miyazaki, "Review shape-memory materials and hybrid composites for smart systems," *Journal of Materials Science*, vol. 33, no. 15, pp. 3743–3762, 1998.

[2] A. D. Moghadam, B. F. Schultz, J. B. Ferguson, E. Omrani, P. K. Rohatgi, and N. Gupta, "Functional metal matrix composites: Self-lubricating, self-healing, and nanocomposites – An outlook," *JOM*, vol. 66, no. 6, pp. 872–881, 2014, doi:10.1007/s11837-014-0948-5

[3] S. X.-D. Karamichailidou and G. Haidemenopoulos, "The unique properties, manufacturing processes and applications of near equatomic Ni-Ti alloys," University of Thessaly Department of Mechanical Engineering Laboratory of Materials, 2016.

[4] W. Hoyer and J. Gasser, "Surface tension and density of binary lead and lead-free Sn-based solders," *Journal of Physics. Condensed Matter*, vol. 17, no. 50, 2005, doi:10.1088/0953-8984/17/50/007

[5] D. Y. Wu, S. Meure, and D. Solomon, "Self-healing polymeric materials: A review of recent developments," May 2008, doi:10.1016/j.progpolymsci.2008.02.001

[6] K. Cerdan et al., "Sustainability of self-healing polymers: A holistic perspective towards circularity in polymer networks," May 1, 2024. Elsevier Ltd, doi:10.1016/j.progpolymsci.2024.101816

[7] R. Purohit, N. K. Gupta, M. R. Purohit, A. Patil, R. K. Bharilya, and S. K. Singh, "An investigation on manufacturing of self-healing materials," *Materials Today: Proceedings*, 2015, pp. 3371–3377. Elsevier Ltd, doi:10.1016/j.matpr.2015.07.311

[8] M. Kleiner, M. Geiger, and A. Klaus. Manufacturing of lightweight components by metal forming. *CIRP Annals - Manufacturing Technology*, vol. 52, no. 2, pp. 521–542, 2003, https://doi.org/10.1016/S0007-8506(07)60202-9

[9] J. F. V. Vincent, O. A. Bogatyreva, N. R. Bogatyrev, A. Bowyer, and A. K. Pahl, "Biomimetics: Its practice and theory," Aug. 22, 2006, Royal Society, doi:10.1098/rsif.2006.0127

[10] P. Nnamchi, A. Younes, and S. González, "A review on shape memory metallic alloys and their critical stress for twinning." *Intermetallics*, vol. 105, pp. 61–78.

[11] R. P. Wool and K. M. O'Connor, "A theory crack healing in polymers," *Journal of Applied Physics*, vol. 52, no. 10, pp. 5953–5963, Oct. 1981, doi:10.1063/1.328526

[12] H. Laieghi, S. Alipour, and A. Mostafapour, "Heat-assisted friction stir welding of polymeric nanocomposite," *Science and Technology of Welding and Joining*, vol. 25, no. 1, pp. 56–65, 2020, doi:10.1080/13621718.2019.1610613

[13] S. Sharma, G. Nandan, P. Rohatgi, and R. Prakash, "Recent advances in self-healing materials," *Materials Today Proceedings*, vol. 18, pp. 4729–4737, 2019, doi:10.1016/j.matpr.2019.07.460

[14] Y. C. Yuan, T. Yin, M. Z. Rong, and M. Q. Zhang, "Self healing in polymers and polymer composites. Concepts, realization and outlook: A review," *Express Polymer Letters*, vol. 2, no. 4, pp. 238–250, 2008, doi:10.3144/expresspolymlett.2008.29

[15] S. Sharma and G. Nandan, "Casting and comprehensive analysis of nitinol-based self-healing metal matrix composite of A356 alloy prepared via clamp technique," *Engineering Research Express*, vol. 6, no. 2, p. 025568, Jun. 2024, doi:10.1088/2631-8695/ad52ea

[16] S. Sharma, M. Bellah, V. Srivastava, and P. Rohatgi, "Publisher correction: Gravity and semisolid casting of self-healing aluminum matrix composites," *International Journal of Metalcasting*, May 2024, doi:10.1007/s40962-024-01364-y

[17] M. Nosonovsky and P. K. Rohatgi, "Development of metallic and metal matrix composite self-healing materials," in *Biomimetics in Material Science*, R. Hull, C. Jagadish, R. M. Osgood, Jr. J. Parisi, and Z. Wang, Eds., New York: Springer, 2012, ch. 5, pp. 87–122.

[18] G. Lindwall et al., "Diffusion in the Ti-Al-V System," *Journal of Phase Equilibria and Diffusion*, vol. 39, no. 5, pp. 731–746, Oct. 2018, doi:10.1007/s11669-018-0673-9
[19] T. Basner, "Rheocasting of semi-solid A357 aluminum," *SAE Technical Papers*, no. 724, 2000, doi:10.4271/2000-01-0059
[20] L. Pernigoni, U. Lafont, and A. M. Grande, "Self-healing materials for space applications: Overview of present development and major limitations," *CEAS Space Journal*, vol. 13, no. 3, pp. 341–352, Jul. 2021, doi:10.1007/s12567-021-00365-5
[21] F. Farshi Azhar and A. Ahmadinia, "Self-healing cementitious materials containing encapsulated epoxy-montmorillonite-calcium nitrate: Evaluating crack-healing performance, mechanical and thermal properties," Journal of Sustainable Cement-Based Materials, vol. 12, no. 1, pp. 36–48, Jan. 2023, doi:10.1080/21650373.2021.2012724
[22] O. Lashkari, S. Nafisi, J. Langlais, and R. Ghomashchi, "The effect of partial decanting on the chemical composition of semi-solid hypo-eutectic Al-Si alloys during solidification," *Journal of Materials Processing Technology*, vol. 182, no. 1–3, pp. 95–100, Feb. 2007, doi:10.1016/j.jmatprotec.2006.06.021
[23] S. Utrera-Barrios, R. Verdejo, M. Á. López-Manchado, and M. Hernández Santana, "Self-healing elastomers: A sustainable solution for automotive applications," May 25, 2023, doi:10.1016/j.eurpolymj.2023.112023
[24] S. Kim, W. Goo, G. Karima, J. H. Lee, and H. D. Kim, "Polyacrylamide/gel-based self-healing artificial tympanic membrane for drug delivery of otitis treatment," *Biomaterials Research*, vol. 28, 2024, doi:10.34133/bmr.0049
[25] K. Choi et al., "Highly flexible and transparent self-healing elastomer based on localized supramolecular interactions with enhanced mechanical properties and long-term storage stability," *Polymer Degradation and Stability*, vol. 225, p. 110813, Jul. 2024, doi:10.1016/j.polymdegradstab.2024.110813
[26] E. Kilickap, "Expert systems with applications optimization of cutting parameters on delamination based on Taguchi method during drilling of GFRP composite," *Expert Systems with Applications*, vol. 37, no. 8, pp. 6116–6122, 2010, doi:10.1016/j.eswa.2010.02.023
[27] V. Kilicli, X. Yan, N. Salowitz, and P. K. Rohatgi, "Recent advancements in self-healing metallic materials and self-healing metal matrix composites," *JOM*, vol. 70, no. 6, pp. 846–854, 2018, doi:10.1007/s11837-018-2835-y
[28] M. L. Durão, L. Nobre, C. Mota, J. Bessa, F. Cunha, and R. Fangueiro, "Self-healing composites: A path to redefining material resilience—A comprehensive recent review," *Materials*, vol. 17, no. 19, p. 4681, Sep. 2024, doi:10.3390/ma17194681
[29] R. Ikura, J. Park, M. Osaki, H. Yamaguchi, A. Harada, and Y. Takashima, "Design of self-healing and self-restoring materials utilizing reversible and movable crosslinks," *Nature Publishing Group Asia Materials*, vol. 14, no. 1, p. 10, Dec. 2022, doi:10.1038/s41427-021-00349-1
[30] L. Ivanchev, D. Wilkins, G. Govender, W. Du Preez, and R. Bean, "Rheo-processing of semi-solid metal alloys: A new technology for manufacturing automotive and aerospace components," *South African Journal of Science*, vol. 104, no. 7–8, pp. 257–259, 2008.
[31] A. K. Chattopadhyay and M. R. Zentner, "Aerospace and aircraft coatings," in *Federation Series on Coatings Technology*, Philadelphia, 1990, 32.
[32] X. G. Zheng, Y. Shi, and K. Lu, "Electro-healing cracks in nickel," *Materials Science and Engineering A*, vol. 561, pp. 52–59, 2013, doi:10.1016/j.msea.2012.10.080
[33] V. Giurgiutiu, "Electro-mechanical impedance method for crack detection in thin wall structures," In *3rd Int. Workshop of Structural Health Monitoring*, 2001 (pp. 12–14).
[34] Q. Shengru, "Damage healing of aluminum alloys by D. C. electropulsing and evaluation by resistance," *Rare Metal Materials and Engineering*, vol. 38, no. 4, pp. 570–573, 2024, doi:10.1016/S1875-5372(10)60027-6

[35] W. Aiping, Z. Guisheng, and R. Jialie, "Formation process of the bonding joint in Ti / Al diffusion bonding," *Materials Science and Engineering A*, vol. 480, pp. 456–463, 2008, doi:10.1016/j.msea.2007.07.027
[36] R. Jiangwei, L. Yajiang, and F. Tao, "Microstructure characteristics in the interface zone of Ti / Al diffusion bonding," *Materials Letters*, vol. 56, no. November, pp. 647–652, 2002.
[37] Y. Zhang, Z. Wei, L. Shi, and M. Xi, "Characterization of laser powder deposited Ti-TiC composites and functional gradient materials," *Journal of Materials Processing Technology*, vol. 206, no. 1–3, pp. 438–444, 2008, doi:10.1016/j.jmatprotec.2007.12.055
[38] H. Ebrahimnezhad-Khaljiri and R. Eslami-Farsani, "Experimental investigation of flexural properties of glass fiber–epoxy self-healable composite structures containing capsulated epoxy healing agent and NiCl2(imidazole)4 catalyst," *Journal of Industrial Textiles*, vol. 51, no. 5, pp. 788–805, 2021, doi:10.1177/1528083719892923
[39] J. Park, J.-M. Heo, S. Seong, J. Noh, and J.-M. Kim, "Self-assembly using a retro Diels-Alder reaction," *Nature Communications*, vol. 12, no. 1, p. 4207, Jul. 2021, doi:10.1038/s41467-021-24492-z
[40] M. Bellah, M. Nosonovsky, and P. Rohatgi, "Shape memory alloy reinforced self-healing metal matrix composites," *Applied Sciences*, vol. 13, no. 12, 2023, doi:10.3390/app13126884
[41] X. Luo and P. T. Mather, "Shape memory assisted self-healing coating," *ACS Macro Letters*, vol. 2, no. 2, pp. 152–156, 2013, doi:10.1021/mz400017x
[42] S. van der Zwaag, N. H. van Dijk, H. M. Jonkers, S. D. Mookhoek, and W. G. Sloof, "Self-healing behaviour in man-made engineering materials: Bioinspired but taking into account their intrinsic character," *Philosophical Transactions of the Royal Society A - Mathematical Physical and Engineering Sciences*, vol. 367, no. 1894, pp. 1689–1704, May 2009, doi:10.1098/rsta.2009.0020
[43] V. Srivastava and M. Gupta, "Impact of post hardening mechanism on self-healing assessment of AA2014 nitinol-based smart composites," *Metals and Materials International*, vol. 27, no. 8, pp. 2666–2681, 2021, doi:10.1007/s12540-020-00630-y
[44] M. Nosonovsky and P. K. Rohatgi, *Biomimetics in Materials Science: Self-Healing, Self-Lubricating, and Self-Cleaning Materials*, vol. 152, Springer Series in Materials Science. New York: Springer, 2012, doi:10.1007/978-1-4614-0926-7
[45] J. M. Lucci, R. S. Amano, P. Rohatgi, and B. Schultz, "Self-healing in an aluminum alloy reinforced with microtubes," in *2008 Proceedings of 3rd Energy Nanotechnology International Conference, ENIC 2008*, pp. 79–88, 2009, doi:10.1115/ENIC2008-53011
[46] R. N. Lumley and I. J. Polmear, "Advances in self healing of metals," in *Proceedings of the First International Conference on Self Healings*, 2007, vol. 100, pp. 1–11. Springer.
[47] G. R. Johnson and W. H. Cook, "A constitutive model and data for metals subjected to large strains, high strain rates and high temperatures." *Proceedings 7th International Symposium on Ballistics*, The Hague, 19–21 April 1983, 541–547.
[48] N. K. Gupta, G. D. Thakre, and M. Kumar, "Dry sliding wear performance on self-healing Al6061 composites," *Journal of Engineering, Design and Technology*, vol. 18, no. 5, pp. 1357–1370, Jan. 2020, doi:10.1108/JEDT-03-2020-0078
[49] G. Parande, V. Manakari, S. Wakeel, M. S. Kujur, and M. Gupta, "Enhancing mechanical response of monolithic magnesium using nano-NiTi (Nitinol) particles," *Metals (Basel)*, vol. 8, no. 12, pp. 1–13, 2018, doi:10.3390/met8121014
[50] R. Kumar Pittala, G. Dhanaraju, B. Satish Ben, and B. Avinash Ben, "Self-healing of matrix cracking and delamination damage assessment in microcapsules reinforced carbon fibre epoxy composite under flexural loading," *Composite Structures*, vol. 291, p. 115691, Jul. 2022, doi:10.1016/J.COMPSTRUCT.2022.115691

[51] C. S. Paglia, K. V. Jata, and R. G. Buchheit, "A cast 7050 friction stir weld with scandium: Microstructure, corrosion and environmental assisted cracking," *Materials Science and Engineering A*, vol. 424, no. 1–2, pp. 196–204, 2006, doi:10.1016/j.msea.2006.03.065

[52] P. Longo et al., "Development of a new stable ruthenium initiator suitably designed for self-repairing applications in high reactive environments," *Journal of Industrial and Engineering Chemistry*, vol. 54, pp. 234–251, 2017, doi:10.1016/j.jiec.2017.05.038

[53] R. Ahmad and M. B. A. Asmael, "Effect of aging time on microstructure and mechanical properties of AA6061 friction stir welding joints," *International Journal of Automotive and Mechanical Engineering*, vol. 11, no. 1, pp. 2364–2372, 2015, doi:10.15282/ijame.11.2015.17.0198

6 Applications of Self-Healing Materials in Energy and Electronics

T. Ravi, Falguni Tlajiya, Awakash Mishra, V. Haripriya, Chanchal Ahlawat, Anant Bhardwaj, and Mehran Khan Alam

6.1 INTRODUCTION

Self-healing materials that are integrated into energy storage systems, specifically batteries, have created a groundbreaking advancement in energy storage fields and technology. Material degradation and failure in batteries is a growing problem because the demand keeps increasing for durable, high-performance, long-lasting storage solutions, particularly in electric vehicles, along with portable electronic gadgets and grid storage operations (Ma et al., 2021). Self-healing materials provide an innovative response through their ability to conduct independent self-repair of damaged battery components, which both increases device operational duration and maintains system performance, while enhancing safety (Meng et al., 2024). The cycling process of lithium-ion batteries produces microcracks at electrode interfaces and electrolyte junctions that trigger performance reductions until thermal catastrophic events occur (Marinow et al., 2023). SH batteries, which incorporate responsive polymers, can automatically repair injuries on the structural and electrochemical levels through physical or chemical mechanisms to preserve performance and durability over time (Song et al., 2025). Figure 6.1 shows the multifunctional aspects of SH battery materials.

Scientists have investigated numerous healing mechanisms, starting from reversible chemical bonds along with capsule-based healing agents and supramolecular interactions, which have been designed specifically for individual battery components such as anodes, cathodes and separators (Vegge et al., 2021). SH techniques are essential components for flexible and wearable electronic devices since they protect against mechanical breakdown caused by bending and stretching (Tan et al., 2024). TPS self-healing electrolytes and binders ensure performance quality in such devices (Zhu et al., 2023). Renewable energy storage systems equipped with SH systems lower maintenance expenses and enhance their environmental sustainability (Bertaglia et al., 2024). SH materials extend their durability in ways that drive developments toward more reliable energy storage technologies with greater efficiency, which support both

DOI: 10.1201/9781003639510-6

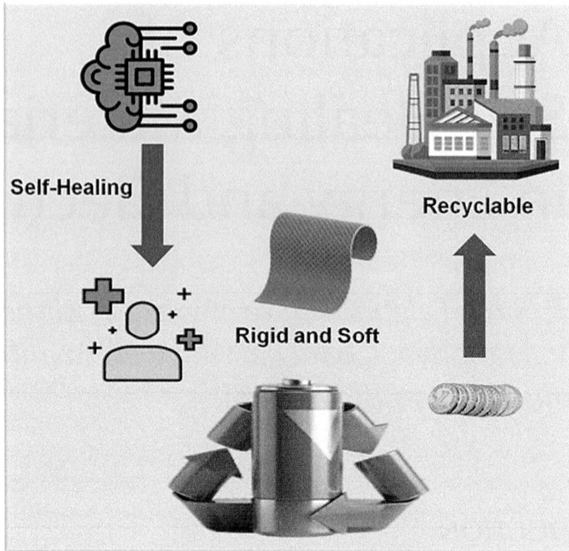

FIGURE 6.1 Multifunctional aspects of SH battery materials.

consumer electronics and large-scale systems. Solar and wind energy storage applications will benefit substantially from SH materials since these technologies will lead to the development of sustainable energy storage systems.

6.2 EXPLORING SH MATERIALS FOR ENHANCED SOLAR PANEL AND ENERGY STORAGE EFFICIENCY

SH materials function as key elements to extend both the durability and operational capacity of solar panels and energy storage structures. Different materials used in solar cells feature SH properties, which help fix photovoltaic layer micro-fractures from thermal cycling and mechanical stress that would deteriorate efficiency rates throughout time (Zhang et al., 2025). Protecting the reliability of solar panels across different environments depends heavily on this function; such materials at present appear in batteries and supercapacitors to recover functionality and structural durability within damaged electrolytes and electrodes.

Through its automatic self-repair process, the device independently sustains its operational efficiency, thus retaining a longer lifetime without expensive upkeep needs. SH technologies benefit sustainability by improving renewable energy systems both through efficiency and reliability increases and through reduction of environmental effects from device end-of-life. It becomes essential to have self-repair capabilities in tough environments that cause components to break more regularly. SH materials will become indispensable for long-term renewable energy solutions due to growing market demand because they will maximize present and future energy technology performance and sustainability. Table 6.1 shows the comparison of SH energy materials.

TABLE 6.1
Comparison of SH Energy Materials

Aspect	Solar Panels	Energy Storage
Primary function	Maintain photovoltaic efficiency and longevity	Maintain electrochemical performance and cycle life
Common materials used	SH polymers, elastomers, hydrogels	SH binders, solid-state electrolytes, conductive gels
SH mechanism	UV/heat-induced, reversible bonding, microencapsulation	Reversible covalent bonding, ionic interactions, capsule-based
Targeted components	Protective coatings, encapsulants, substrates	Electrodes (anode/cathode), electrolytes, separators
Damage type addressed	Surface cracks, delamination, moisture ingress	Microcracks, dendrite formation, interface breakdown
Benefits	Enhanced weather resistance, longer lifespan	Improved safety, stability, and performance over cycles
Challenges	Environmental exposure, integration with transparent layers	Electrochemical stability, scalability, and conductivity issues
Typical applications	Outdoor solar modules, flexible solar panels	Lithium-ion, solid-state, and flexible batteries

6.2.1 SH Materials in Solar Panels

SH materials appear as an effective approach to boost solar power system durability while improving performance efficiency. UV radiation and temperature changes alongside mechanical stresses affect solar panels throughout their operational lifespan (Sabet, 2024). Photovoltaic layers and encapsulants featuring SH polymers or coatings can detect and repair minimal surface damage as well as small cracks and delamination without requiring external intervention. Solar panel performance remains optimal since the built-in self-repair mechanism sustainably maintains electrical conductivity in conjunction with light transmission capabilities (Xu et al., 2023). Solar energy technology benefits from SH materials because they lengthen the operational lifetime of PV systems, lower maintenance expenses, and enhance power production at the same time. The innovative material contributes to the long-term operation and affordability of solar applications, thus improving solar energy systems' reliability as sustainable alternatives.

6.2.2 SH Materials in Energy Storage

Energy storage systems and their batteries experience enhanced durability and reliability on implementing SH materials. The cycling and mechanical tension applied to lithium-ion along with next-generation batteries triggers electrode and electrolyte deterioration through microcracking (Davino et al., 2022). SH polymers and microcapsules used within the materials enable autonomous damage repair, thus supporting structural stability while lengthening battery service life. This technology operates to enhance both performance levels and safety and assists in creating flexible and

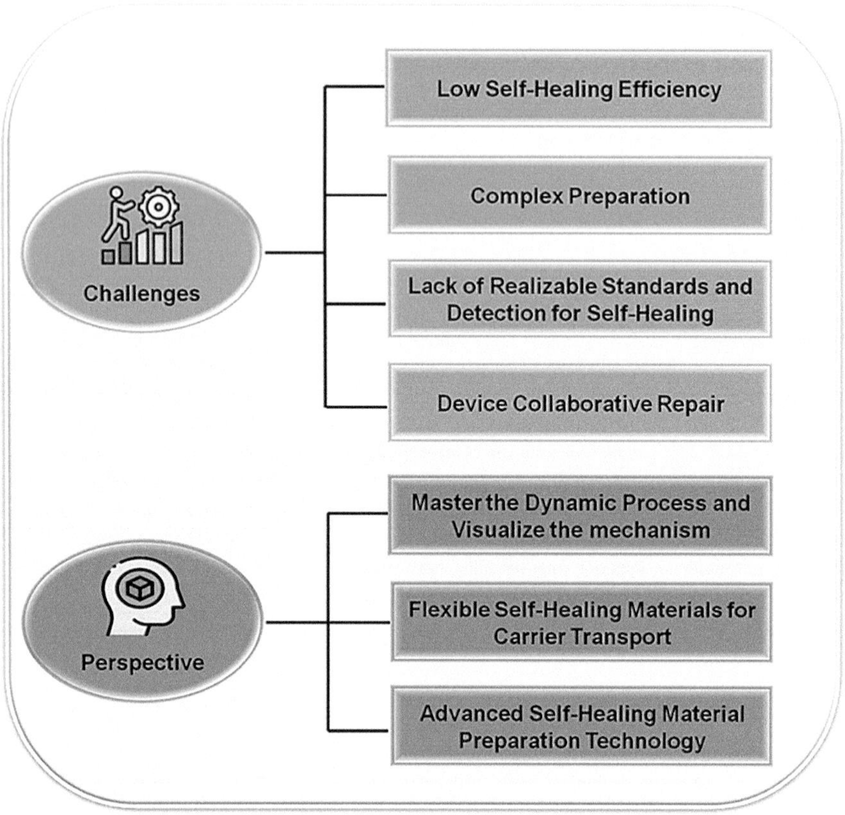

FIGURE 6.2 Challenges and perspectives for self-healing energy storage.

wearable energy devices. SH mechanisms offer an exciting strategy to develop sustainable energy storage systems with better resilience. Figure 6.2 displays the challenges and perspectives for SH energy storage.

6.2.3 1D Nanomaterials in SH Batteries

One-dimensional (1D) nanomaterials are used for developing SH electrodes, as they improve battery and supercapacitor stability and cycle ability (Ibrahim et al., 2025). The physical repair of circuits depends on flexible materials, which maintain excellent conductivity in 1D nanomaterials such as nanowires, nanotubes, and nanofibers. Scientists have developed the first SH LIB structure by utilizing gel electrolytes along with sodium carboxymethylcellulose and lithium sulfate-containing membranes (Li et al., 2022a, 2022b). Enhancing the Durability of Batteries and Electronic Devices

SH materials significantly improve the durability and performance of batteries by autonomously repairing damage to critical components like electrodes and separators, which typically degrade over time due to mechanical stress or cycling. The incorporation of SH polymers within lithium-ion batteries works to solve electrode capacity loss from microcracks caused by usage breakdown. Through this repair capability, the

TABLE 6.2
Comparison of Traditional and SH Materials

Parameter	Traditional Materials	SH Materials
Durability	Degrades over time due to mechanical and chemical stress	Maintains integrity by autonomously repairing micro-damage
Battery lifespan	Shorter; limited by electrode degradation and electrolyte breakdown	Extended; cracks and failures are repaired, preserving performance
Mechanical stability	Susceptible to cracking, delamination, and wear	Improved resilience to mechanical deformation and cycling fatigue
Safety	Increased risk of thermal runaway and leakage after damage	Reduced failure risk due to self-repair mechanisms
Environmental impact	Higher e-waste generation due to frequent replacements	Lower e-waste due to longer component life
Cost-efficiency (long-term)	Higher maintenance and replacement costs	Reduced lifecycle cost despite higher initial material costs
Application in flexible devices	Limited flexibility; prone to failure on bending/stretching	Ideal for wearables and bendable devices
Technology readiness level (TRL)	Commercially mature and widely adopted	Emerging; under active research and development

battery remains efficient, extends its lifespan, and improves its performance qualities (Zhao & Sottos, 2021). SH technology delivers its best advantage to battery-dependent industries such as electric vehicles and portable electronics because battery performance and safety remain crucial. These materials can transform battery science because they both decrease replacement requirements and make battery components more durable. These materials enhance battery performance and lead to environmental sustainability because they help decrease electronic waste while reducing the environmental effect of waste disposal. SH technology advancements establish the foundation for building dependable and enduring energy storage systems that propel sustainable and efficient energy system development. Table 6.2 illustrates the comparison of traditional and SH materials.

6.2.4 Enhancing Durability of Batteries

SH materials deliver elevated durability to batteries by solving problems linked to material breakdown and electrode fractures together with electrolyte loss rates. Regular cycling operations cause damage to fundamental system elements in lithium-ion batteries (Malik et al., 2023). Batteries containing embedded SH materials or polymers possess the ability to heal microcracks automatically and thus maintain their structural strength, thereby prolonging their operational period. This not only improves the longevity of energy storage systems but also ensures consistent performance, reducing the frequency of replacements and maintenance (Roy et al., 2024). Such advancements contribute to the development of more sustainable, reliable, and cost-effective energy storage solutions for both consumer electronics and electric vehicles.

6.2.5 Enhancing the Durability of Electronic Devices

The durability of electronic devices experiences significant improvement through SH materials, specifically in energy storage systems, which involve batteries. The continuous charge-discharge process of batteries will gradually cause structural breakdown, which results in decreased performance or failed operations. Incorporated SH polymers or composites enable batteries to perform automatic repairs on electrodes and electrolytes at the points of crack and wound and chemical damage, thereby prolonging operational lifetime while enhancing reliability (Chen et al., 2022). Through this innovation, energy storage systems achieve better sustainability because they minimize critical failures, such as thermal runaway and reduce total maintenance costs while increasing their operational life span. Figure 6.3 shows the SH hydrogels for durable electronics.

6.2.6 Biosourced SH Materials for Battery Durability

Based on their molecular makeup, bio-sourced polymers can be divided into three main groups: polyesters, polysaccharides, and proteins. Obtainable from renewable sources such as cellulose, polylactide, and lignin, polysaccharides can be copolymerized into gel polymer electrolytes or utilized as a binder (Narayan et al., 2022). Natural polymers provide a wealth of modification options, including the ability to support battery SH capabilities and modify their characteristics. Proteins, along with

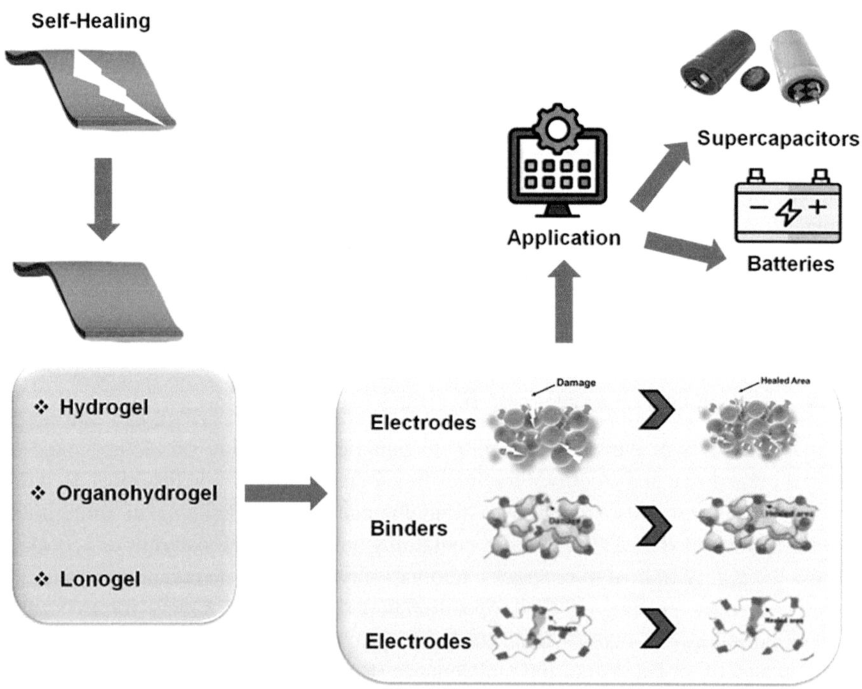

FIGURE 6.3 SH Hydrogels for durable electronics.

TABLE 6.3
SH Efficiency of Biosourced Battery Materials

Material Type	Healing Time (hrs)	Test Cycle (No. of Cycles)	SH Efficiency (%)	Retained Capacity (%) After Healing
Chitosan-based polymer	4	100	87.5	91.2
Cellulose nanofiber composite	6	150	82.3	88.7
Soy protein isolate hydrogel	3	120	90.4	93.1
Lignin-infused polyurethane	5	130	85.6	89.9
Starch-based elastomer	4.5	110	80.2	86.4

lipids and sterols, form specific barriers that prevent ions, molecules, or biomolecules from moving freely in aqueous media. These channels, also known as protein nanopores, have been effectively applied to electrically detect single molecules, tackling issues in biotechnology and health (Lu et al., 2025). Certain biological pores such as lysenin can serve as building blocks for batteries' extremely selective separators. The shape and function of proteins can be altered by altering the amino acid sequence of the pore. Large-scale production of necessary proteins and efficient engineering exploration are made possible by biofactories, which are inexpensive to grow and simple to manage. Table 6.3 shows the SH efficiency of biosourced battery materials.

6.2.7 SH Materials in Lithium-Ion Batteries (LIBs)

LIBs are the primary energy storage devices in electronic devices due to their high density, long lifespan, and environmental friendliness (Gai et al., 2021). It could be damaged or broken due to deformation, especially in flexible wearable electronics. Mechanical microcracks during repetitive charge/discharge cycles can reduce LIBs' cycle stability. SH LIBs can improve cycle stability and lifespan by healing external damage or microcracks.

6.3 REDUCING E-WASTE THROUGH SH TECHNOLOGIES

SH technologies integrated into batteries show great potential to fight against electronic waste (e-waste) buildup. Parallel batteries run into failure problems when electrode cracks and electrolyte break down before their intended service life ends, thus triggering their disposal and adding to massive e-waste accumulation (Enayati-Gerdroodbar et al., 2024). SH polymers and composite additions to battery technology enable automatic reaction to material damage that prolongs a battery's operational time. Battery lifespans lengthen by this enhancement because it reduces replacement frequency, thus decreasing environmental strain from battery production and disposal operations. Electronic devices particularly wearable technologies along with electric vehicles will experience greater adoption when SH battery technologies become mainstream because they ensure sustainability advancement.

TABLE 6.4
Performance Comparison of Battery Waste Reduction

Parameter	Conventional Batteries	SH Batteries
Cycle life (number of cycles)	500–1,000 cycles	1,500–5,000 cycles
Failure rate (%)	10–20% after 1 year	<5% after 1 year
Component replacement frequency	High (due to wear and cracking)	Low (due to autonomous healing)
Material waste generation (g/year)	100–300 g per device	30–100 g per device
Recycling efficiency (%)	30–50% (varies with battery type)	50–70% (improved due to longer lifespan)
Environmental impact index	High	Low
Operational downtime (hours/year)	50–100 hours	<20 hours
Estimated lifespan (years)	2–3 years	5–8 years

These approaches for circular economics work to decrease the expanding e-waste problem worldwide while simultaneously protecting resources and our environment. Batteries equipped with SH materials extend their operational lifespan, allowing consumers and manufacturers to cut costs together with diminished raw material requirements. Sustainable battery technology took an essential milestone when this innovation appeared while supporting international initiatives to manage e-waste and extend electronic device component durability. Table 6.4 shows the performance comparison of battery waste reduction.

6.3.1 Degradation Challenges in LIB: A Root Cause of E-Waste

LIB dominates today's energy storage market due to its high energy density and performance. Structurally, LIBs consist of a cathode, anode, electrolyte, separator, and packaging materials. Among these, the electrode materials are particularly prone to degradation, which significantly shortens battery lifespan and contributes to the growing issue of electronic waste (Hua et al., 2024). Electrode failure primarily stems from three mechanisms: (1) During repeated charging and discharging, the insertion and extraction of ions cause volumetric changes in electrodes, resulting in mechanical damages such as cracking and delamination. (2) Long-term cycling promotes the growth of lithium dendrites on the negative electrode, which compromises efficiency and cycle life. (3) In flexible and wearable devices, mechanical stresses from bending or stretching can damage the electrode structure.

These forms of degradation often render the battery nonfunctional, leading to premature disposal and contributing to the mounting problem of e-waste. SH technologies offer a compelling solution by enabling electrodes to autonomously repair damage at the microstructural level. Decorative foil bonded to aluminum can stretch and recover after being folded, which extends battery useful life, thus minimizing the need for device battery replacement and cutting electronic waste significantly. SH materials show promise for battery development because they enable progress toward sustainable design along with improved device reliability.

6.3.2 Extending Battery Lifespan with SH

Batteries receive extended lifespans through the incorporation of SH materials because these materials protect against damage, which generally occurs when batteries charge and discharge. Several cycles of traditional batteries result in component degradation through electrode separators and electrolytes that eventually produce capacity loss and device failure (He et al., 2024). The capability of using SH materials consisting of polymers and responsive chemical agents allows components to fix microcracks and preserve structural strength while stopping capacity reduction. The continuous self-repair system of modified separators helps extend battery life and decreases the need for frequent replacements. SH technologies implemented within energy storage systems deliver better performance outcomes and help sustainability efforts through reduced e-waste volumes and lowered maintenance expenses.

The comparison between standard and SH batteries shows that SH technology significantly extends battery lifespan. After 500 charge cycles, the standard battery retains 75% capacity, while the SH battery retains 90%. At 900 cycles, the standard battery drops to 50%, compared to 80% for the SH version. By 1500 cycles, the standard battery holds only 12%, whereas the SH battery maintains 58%. Finally, at 1700 cycles, the standard battery reaches 0% capacity, while the SH battery still retains 50%. Overall, SH batteries demonstrate better performance and longevity across all stages of the charge cycle. Figure 6.4 shows the comparison of battery lifespan with and without SH technology.

6.3.3 Minimizing Electronic Waste via Smart Materials

SH materials play a vital role in reducing electronic waste by significantly extending the operational life of batteries. Traditional batteries often fail due to microstructural damage, leading to premature disposal (Song et al., 2024). By incorporating SH

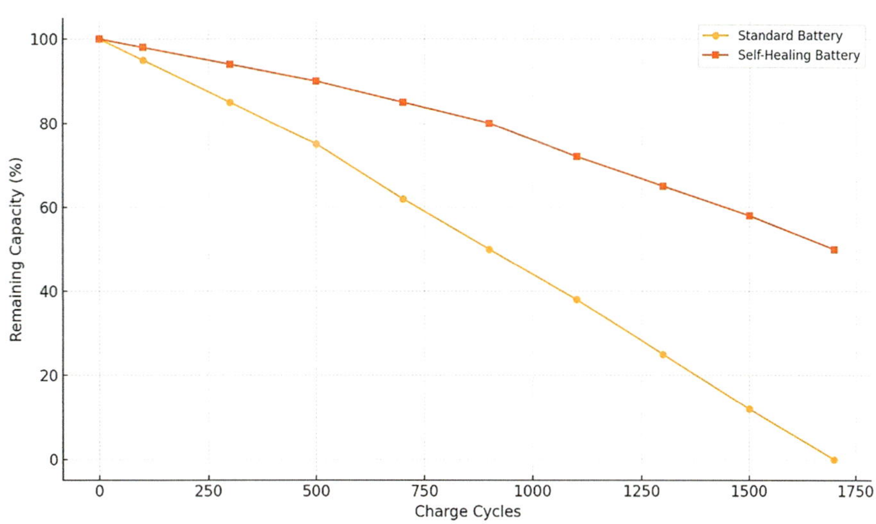

FIGURE 6.4 Comparison of battery lifespan with and without SH technology.

6.4.2 Extrinsic SH Strategies: Microcapsules and Vascular Networks

Self-healing (SH) mechanisms built into battery systems serve two primary functions: repairing damage to restore functionality and extending battery life through dendrite repair and crack prevention. The restoration and strength expansion mechanisms comprise of intrinsic and extrinsic strategies. Microcapsules and vascular systems serve as apparatuses of extrinsic SH by containing healing agents. When the capsules are damaged or the networks are disrupted, they release healing agents through chemical reactions that fill the cracks and restore the structure. This mechanism is particularly effective in battery electrodes and electrolytes, where both electrochemical and mechanical stability are critical. New energy storage systems benefit from these approaches, which improve cycle performance and enhance safety through continued development. Hybrid SH Approaches for Enhanced Battery Durability

SH battery approaches benefit from combination structures between intrinsic and extrinsic elements, which improve performance quality and extend effective lifespan. The material-based repair processes of intrinsic methods use chemical bonds that can spontaneously fix themselves at molecular resolution. The operating mechanism of encapsulated healing agents is to release their content when exposed to damage through external means. The healing processes concentrate on treating battery components, including electrodes, electrolytes, and separators, simultaneously functioning to avoid failures by stopping dendrite growth and temperature-induced damage. Hybrid systems composed of responsive polymer microcapsules and conductive fillers establish mechanisms, which enhance battery cycle durability and maintain safety operations. Strong energy storage systems require a complete strategies overview for developing effective systems that can function optimally in electric vehicles and transferable electronic applications.

6.5 SH SEPARATORS: ENHANCING BATTERY SAFETY

SH extractors function as innovative solutions for battery security features together with enhanced product lifespan in both lithium-ion battery systems and solid-state battery systems. The battery defense system utilizes separators as separation essentials that avoid electronic connection between anode and cathode (Chen et al., 2021). Standard battery separators break down quickly when exposed to two operating factors, which include natural mechanical compression and rising temperatures as well as injury caused by lesions that can cause severe battery breakdowns including thermal runaway and short circuits. This extractor technology comprises detailed recovery systems, which achieve repairs to its mechanical apparatuses.

Specific composites throughout the materials contain reactive polymers that enable realignment across surfaces when unprotected to particular triggers such as temperature changes or electrical fields. Artificial vascular networks together with microcapsules function as SH components because they contain therapeutic managers inside to automatically release this content through breaches after injury occurs while creating seals to stop additional damage. The operational time of battery separators depends on SH features for enhancing their safety presentation while protecting them against damage. The innovative configuration enables batteries to preserve longer operation time through fewer replacements, thus producing better dependability and increased stored energy capacity. Separators that self-heal will significantly improve energy and electronic battery system security and strength.

6.5.1 Advancements in Smart Separator Materials for Preventing Battery Failures

SH separators within batteries service as crucial components because they stop short circuits and thermal runaway events. Failure of the separator unit causes dangerous battery malfunctions because it physically separates the anode from the cathode. Scientific research has targeted smart separator materials to grow automatic damage repair arrangements, which remedy physical breaks stemming from cathode-anode inter-stress or external occurrence damage. The materials incorporate thermally approachable polymers or microencapsulation methods to enable the extractor to repeatedly fix cracks while preserving its structural integrity, thus avoiding additional harm. Improved extractors from contemporary growth association structures enhance the thermal stability and lower heat-related fire risks. SH functionality integrated into separators boosts battery durability as well as reliability and protection feature specifically for electric vehicles and portable electronic applications. The growth of these innovations combines crucial technology to enhance battery functioning and simplifies risk reduction through battery failures.

6.5.2 Mechanisms and Benefits of SH Separators in Lithium-Ion and Solid-State Batteries

SH separators play a critical role in improving the security and durability of lithium-ion and solid-state batteries by avoiding short circuits and mitigating thermal runaway. In these batteries, separators are crucial for preventing direct contact between electrodes while allowing ionic conductivity (Li et al., 2022a, 2022b). However, mechanical stress degradation over time can lead to failures, compromising battery performance and safety. SH separators address these issues by incorporating materials that can autonomously repair themselves when damaged. These materials often use reversible chemical bonds or polymer networks that respond to external stimuli, such as temperature or motorized stress, to restore the separator's integrity. For lithium-ion batteries, this helps prevent interior shorts, while in solid-state batteries, SH separators maintain the stability of the solid electrolyte interface. The result is improved cycle life, enhanced safety, and a reduced risk of catastrophic failure, making SH separators a promising solution for next-generation energy storage systems.

6.6 CONCLUSION

SH materials represent a significant advancement in the pursuit of more durable, efficient, and maintainable energy and automatic schemes. By enabling autonomous repair through sophisticated apparatuses, such as reversible bonding and microencapsulation, these smart materials address key difficulties like material fatigue, mechanical failure, and reduced device lifespan. Their successful integration into solar panels, lithium-ion batteries, and flexible electronics not only enhances performance and dependability but also reduces conservation costs and environmental impact. The growth of novel SH polymers and composites tailored through techniques like nanocomposite reinforcement and in situ polymerization demonstrates the growing maturity of this field. As demand for resilient, long-lasting technologies continues to grow, particularly in renewable energy and user electronics, SH resources offer a forward-looking solution. Ultimately, their ability to extend product lifecycles and minimize electronic waste positions them as a cornerstone in the growth toward greener, smarter, and more sustainable technological systems.

REFERENCES

Bertaglia, T., Costa, C. M., Lanceros-Méndez, S., &Crespilho, F. N. (2024). Eco-friendly, sustainable, and safe energy storage: A nature-inspired materials paradigm shift. *Materials Advances*, 5(19), 7534–7547. https://doi.org/10.1039/D4MA00363B

Chen, J., Wang, L., Xu, X., Liu, G., Liu, H., Qiao, Y., . . . & Wang, T. (2022). Self-healing materials-based electronic skin: Mechanism, development and applications. *Gels*, 8(6), 356. https://doi.org/10.3390/gels8060356

Chen, L., Cai, X., Sun, Z., Zhang, B., Bao, Y., Liu, Z., Han, D., & Niu, L. (2021). Self-healing of a covalently cross-linked polymer electrolyte membrane by Diels-Alder cycloaddition and electrolyte embedding for lithium-ion batteries. *Polymers*, 13(23), 4155. https://doi.org/10.3390/polym13234155

Davino, S., Callegari, D., Pasini, D., Thomas, M., Nicotera, I., Bonizzoni, S., . . . & Quartarone, E. (2022). Cross-linked gel electrolytes with self-healing functionalities for smart lithium batteries. *ACS Applied Materials & Interfaces*, 14(46), 51941–51953. https://doi.org/10.1021/acsami.2c15011

Dong, Y., & Li, J. (2022). Oxide cathodes: Functions, instabilities, self-healing, and degradation mitigations. *Chemical Reviews*, 123(2), 811–833. https://doi.org/10.1021/acs.chemrev.2c00251

Enayati-Gerdroodbar, A., Khayati, A., Zolfagharian, A., Salami-Kalajahi, M., &Bodaghi, M. (2024). Vitrimers for sustainable electronics and control of e-waste. *Smart Materials & Methods*, 1(3), 251–293. https://doi.org/10.1080/29963176.2024.2417225

Gai, Y., Li, H., & Li, Z. (2021). Self-healing functional electronic devices. *Small*, 17(41), 2101383. https://doi.org/10.1002/smll.202101383

He, S., Huang, S., Liu, X., Zeng, X., Chen, H., Zhao, L., Noor, H., & Hou, X. (2024). Interfacial self-healing polymer electrolytes for long-cycle silicon anodes in high-performance solid-state lithium batteries. *Journal of Colloid and Interface Science*, 665, 299–312. https://doi.org/10.1016/j.jcis.2024.03.118

Hua, C., Gao, J., & Liu, J. (2024). Room temperature self-healing liquid metals: Capabilities, applications and challenges. *International Journal of Smart and Nano Materials*, 15(3), 469–501. https://doi.org/10.1080/19475411.2024.2385349

Ibrahim, O. O., Liu, C., Zhou, S., Jin, B., He, Z., Zhao, W., . . . & Zhang, S. (2025). Recent advances in nanomaterial-based self-healing electrodes towards sensing and energy storage applications. *Sensors*, 25(7), 2248. https://doi.org/10.3390/s25072248

Li, J., Yang, L., Zhang, H., & Ji, X. (2022a). Self-healing composite solid electrolytes with enhanced Li^+ transport and mechanical properties for safe lithium metal batteries. *Chemical Engineering Journal*, 438, 135418. https://doi.org/10.1016/j.cej.2022.135418

Li, L., Zhang, Q., He, B., Pan, R., Wang, Z., Chen, M., . . . & Sun, L. (2022b). Advanced multifunctional aqueous rechargeable batteries design: From materials and devices to systems. *Advanced Materials*, 34(5), 2104327. https://doi.org/10.1002/adma.202104327

Lu, J., Chen, Y., Lei, Y., Jaumaux, P., Tian, H., & Wang, G. (2025). Quasi-solid gel electrolytes for alkali metal battery applications. *Nano-Micro Letters*, 17(1), 1–66. https://doi.org/10.1007/s40820-024-01632-w

Ma, J., Li, Y., Grundish, N. S., Goodenough, J. B., Chen, Y., Guo, L., ... & Wang, C. A. (2021). The 2021 battery technology roadmap. *Journal of Physics D: Applied Physics*, 54(18), 183001. https://doi.org/10.1088/1361-6463/abd353

Malik, Y. T., Shin, S. Y., Jang, J. I., Kim, H. M., Cho, S., Do, Y. R., & Jeon, J. W. (2023). Self-repairable silicon anodes using a multifunctional binder for high-performance lithium-ion batteries. *Small*, 19(9), 2206141. https://doi.org/10.1002/smll.202206141

Marinow, A., Katcharava, Z., & Binder, W. H. (2023). Self-healing polymer electrolytes for next-generation lithium batteries. *Polymers*, 15(5), 1145. https://doi.org/10.3390/polym15051145

Meng, Q., Huang, Y., Li, L., Wu, F., & Chen, R. (2024). Smart batteries for powering the future. *Joule*, 8(2), 344–373. https://doi.org/10.1016/j.joule.2024.01.011

Narayan, R., Laberty-Robert, C., Pelta, J., Tarascon, J. M., & Dominko, R. (2022). Self-healing: An emerging technology for next-generation smart batteries. *Advanced Energy Materials*, 12(17), 2102652. https://doi.org/10.1002/aenm.202102652

Roy, P. M., Sawant, H. H., Shelar, P. P., Sarode, P. U., & Gawande, S. H. (2024). Battery health management perspective of design, optimization, manufacturing, fault detection, and recycling. *Energy Storage and Saving*. https://doi.org/10.1016/j.enss.2024.04.001

Sabet, M. (2024). Unveiling advanced self-healing mechanisms in graphene polymer composites for next-generation applications in aerospace, automotive, and electronics. *Polymer-Plastics Technology and Materials*, 63(15), 2032–2059. https://doi.org/10.1080/25740881.2024.2365282

Song, H. S., Rumon, M. M. H., Rahman Khan, M. M., & Jeong, J. H. (2025). Toward intelligent materials with the promise of self-healing hydrogels in flexible devices. *Polymers*, 17(4), 542. https://doi.org/10.3390/polym17040542

Song, Z., Li, W., Gao, Z., Chen, Y., Wang, D., & Chen, S. (2024). Bio-inspired electrodes with rational spatiotemporal management for lithium-ion batteries. *Advanced Science*, 11(28), 2400405. https://doi.org/10.1002/advs.202400405

Tan, X., Chu, K., Chen, Z., Han, N., Zhang, X., Pan, H., ... & Song, H. (2024). Recent advances in self-healing hydrogel composites for flexible wearable electronic devices. *Nano Research Energy*, 3(3). https://doi.org/10.26599/NRE.2024.9120123

Vegge, T., Tarascon, J. M., & Edström, K. (2021). Toward better and smarter batteries by combining AI with multisensory and self-healing approaches. *Advanced Energy Materials*, 11(23), 2100362. https://doi.org/10.1002/aenm.202100362

Xu, P., Liu, J., Wang, S., Chen, J., Han, B., Meng, Y., ... & Ge, Z. (2023). Dynamic covalent polymer engineering for stable and self-healing perovskite solar cells. *Materials Horizons*, 10(11), 5223–5234. https://doi.org/10.1039/D3MH01293J

Zhang, Q., Soham, D., Liang, Z., & Wan, J. (2025). Advances in wearable energy storage and harvesting systems. *Med-X*, 3(1), 3. https://doi.org/10.1007/s44258-024-00048-w

Zhao, L., &Sottos, N. R. (2021). Autonomous strategies for improved performance and reliability of Li-ion batteries. *Advanced Energy Materials*, 11(5), 2003139. https://doi.org/10.1002/aenm.202003139

Zhu, S., Liu, Z., Li, W., Zhang, H., Dai, G., & Zhou, X. (2023). Research progress of self-healing polymer materials for flexible electronic devices. *Journal of Polymer Science*, 61(15), 1554–1571. https://doi.org/10.1002/pol.20230020

7 Environmental Impact and Future Challenges of Self-Healing Materials

Vaibhav Srivastava, Masum Bellah, and Pradeep Rohatgi

7.1 INTRODUCTION

Self-healing materials have emerged as a new generation of smart materials capable of autonomously repairing damage in the structure, leading to enhanced durability, safety, and lower periodic maintenance during their functionality. As global attention shifts toward sustainable development and environmentally responsible technologies, it is important to assess the ecological footprint and future scalability of self-healing materials. This chapter will explore the environmental implications of using self-healing materials and their upcoming challenges and research directions that will shape the future of this field.

Global discourse increasingly emphasizes the need to reduce carbon emissions through material recycling, whether in plastics, concrete, or metals. Manufacturing new components and structures often requires substantial energy and high investment, resulting in elevated production costs and significant environmental impacts. Additionally, the disposal of waste materials contributes to landfill accumulation, adversely affecting soil quality and ecosystem health. The intergovernmental panel on climate change (IPCC) has emphasized the urgency of reducing global greenhouse gas emissions. The 2023 IPCC synthesis report highlights the need for a 43% reduction in emission by 2030, compared to 2019 level to limit global warming to 1.5°C.[1] Concrete—the most widely used construction material globally—accounts for approximately 9% of global CO_2 emissions. This is primarily due to the calcination of natural limestone and clay in cement manufacturing process, and cement production requires heating materials to around 1450°C, a process that typically relies on the combustion of fossil fuels such as coal, petroleum, and natural gas.[2] Plastic waste can be found throughout oceans, rivers, soils, and even in freshwater streams used for human consumption in the form of microplastics. It is estimated that, following the current trends, about 12,000 Mt of plastic waste will be discarded in landfills or the natural environment by 2025, while nowadays only an estimated amount of 9% of the total manufactured plastics are being recycled.[3] Thermoplastic can be recycled, but the thermosets and ionomers are crosslinked with irreversible

DOI: 10.1201/9781003639510-7

bonds that cannot be recycled. With the rapidly growing electronics market, battery waste has become a significant environmental concern. Lithium-ion batteries, which contain hazardous elements such as lithium, cobalt, and nickel, can leach toxic substances into the soil and contaminate groundwater when improperly disposed of in landfills.[4]

Self-healing materials possess the intrinsic ability to autonomously repair damage, such as cracks resulting from mechanical stress or catastrophic failure. In response to these challenges, the market for self-healing materials has witnessed notable growth over the past decade. With the global self-healing materials market valued at $1.94 billion in 2023 and projected to grow at a CAGR of 23.5%, reaching approximately $8.85 billion by 2030.[5] This research could accelerate market adoption, potentially achieving targets ahead of 2030. The aerospace and defense sector, a major contributor to this growth, could see up to 20% reductions in lifecycle costs by integrating self-healing materials into structural components. Self-healing will help reduce the amount of waste generated during use, saving energy and raw materials by reducing the need for replacements and repairs. The primary motivation behind developing these materials is to extend the service life of existing components, thereby reducing the need for replacements. By preventing the early failures (nano- to microscale) in self-healing materials, the damaged structural could be recovered, thereby helps in minimizing and lowering the investment on the energy demands and resource consumption associated with manufacturing new components.[6–10] As far as mechanical strength is emphasized, the application of self-healing materials could be in space stations, aerospace, and offshore structural components. The diverse and broader applications of self-healing beyond their traditional mechanical strength are emphasized in food packing industries, damage reporting, radiation shielding,[11] self-healing paints and commercial coats (Nissan Motor and Bayer Corporates),[12,13] self-healing smart batteries,[4] self-healing coatings, self-repaired windows, shooting targets,[12,14–17] and self-healing bio-concretes and ceramics.[2,18,19]

In general, self-healing mechanisms can be broadly categorized into autonomous (intrinsic) and non-autonomous (extrinsic) types. Intrinsic self-healing occurs without the need for any external stimulus, relying instead on the material's inherent molecular dynamics or reversible chemical bonds to repair damage.[20–24] In contrast, extrinsic self-healing requires an external stimulus—such as heat, light, moisture, or mechanical force—to activate embedded healing agents (e.g., microcapsules or vascular networks) that respond to damage and initiate repair.

Self-healing materials offer promising applications across various domains. However, their broader adoption is often limited by trade-offs between functional properties and performance characteristics. While significant progress has been made in multiple areas, the development and implementation of self-healing capabilities in metals and metal-based composites remain relatively underexplored and less successful. Therefore, this chapter focuses on providing an overview of existing work on metal based self-healing composites, highlighting current challenges, and exploring possible future directions.

7.2 SELF-HEALING IN METALS: MECHANISMS, CLASSIFICATIONS, AND CURRENT ADVANCES

Broadly, self-healing in metallic systems can be classified into two primary categories. The first category encompasses self-healing intrinsic to metals, which utilizes mechanisms such as surface coatings, electro-pulse, electrochemical/deposition repair, eutectic-based healing, to restore damage at nano- to microscale. The second category involves healing in metal matrix composites on macroscale, by leveraging mechanisms like SMA-induced crack closure, and the integration of healing agents via microcapsules or hollow fibers embedded in the matrix.[25,26] Some systems also incorporate nanoscale SMA particles or fibers to enhance both mechanical and healing performance. Metals face various challenges in achieving self-healing properties due to casting complexity, susceptibility to oxidation at elevated temperatures, low diffusion, undesirable thermal mismatch between healing agent and matrix, low atomic mobility or diffusivity at room temperature, poor wettability, poor bonding between reinforced healing agents and matrix, effective load transfer, etc.[21,27–30]

7.2.1 Shape Memory Alloy-Assisted Self-Healing

Shape Memory Alloys (SMAs) are a special class of materials characterized by two distinct properties: the shape memory effect and pseudo-elasticity.[31] In self-healing composites, SMA fibers are embedded longitudinally within the matrix as reinforcement. When the matrix surface experiences damage or fracture due to mechanical loading, the embedded SMA fibers are stretched and subjected to tensile stress. Upon heating above the transformation temperature, the SMA undergoes a phase change from detwinned martensite to austenite, allowing it to recover its original shape and generate compressive stress at the crack interface, thereby closing or narrowing the crack.[21,27,28,32–34] On further heating, the eutectic metal in the matrix melts and flows into the crack to effectively seal it through phenomenon associated with capillary action and surface tension[35–37] as shown in Figure 7.1. Therefore, SMA-reinforced MMC is an effective design for large crack healing from micro- to macro-level damage scale.

In the past researchers worked on a number of alloy systems (Al, Mg, Zn, Sn, or Sn-Bi alloy) to be used as matrix. Manuel[38] utilized Mg alloy containing 5.7 at. % Zn and 2.7 at. % Al as the matrix material, reinforced with 1 vol.% of continuously aligned NiTi SMA wires. To improve bonding at the interface, the SMA wires underwent etching followed by the application of a 5 nm gold coating via sputtering. The composites were damaged under tensile loading which subsequently heated above transition temperature, showing partial crack closure. Similarly using Sn-13at. %Bi alloy as a matrix material reinforced with Ni-Ti fiber considering as a proof-of-concept composite displaying 94% of strength recovery with decrease in the uniform ductility to 66% after heating at 169°C healing temperature[39] as shown in Figure 7.3. Misra[10,33] fabricated self-healing metal matrix composites (SHMMCs) using Sn-20 wt.% Bi and Bi-10% Sn alloys as matrices, incorporating 20 vol.% NiTi wires through pressure infiltration casting. After flexural testing and subsequent cracking, the specimens were subjected to healing at 145°C. The samples show a recovery of bent shape of 1°/sec with complete crack closure as shown in Figure 7.2 (i) and (ii).

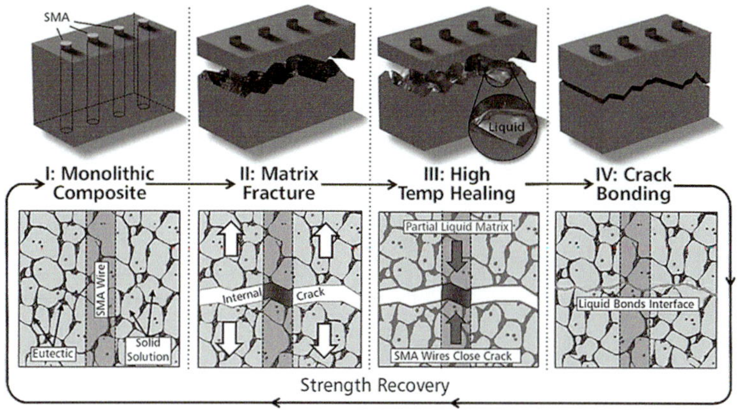

FIGURE 7.1 A schematic representation of mechanism involved for crack narrowing due to SMA reinforced fiber and liquid assistance healing in metal matrix composites.[8]

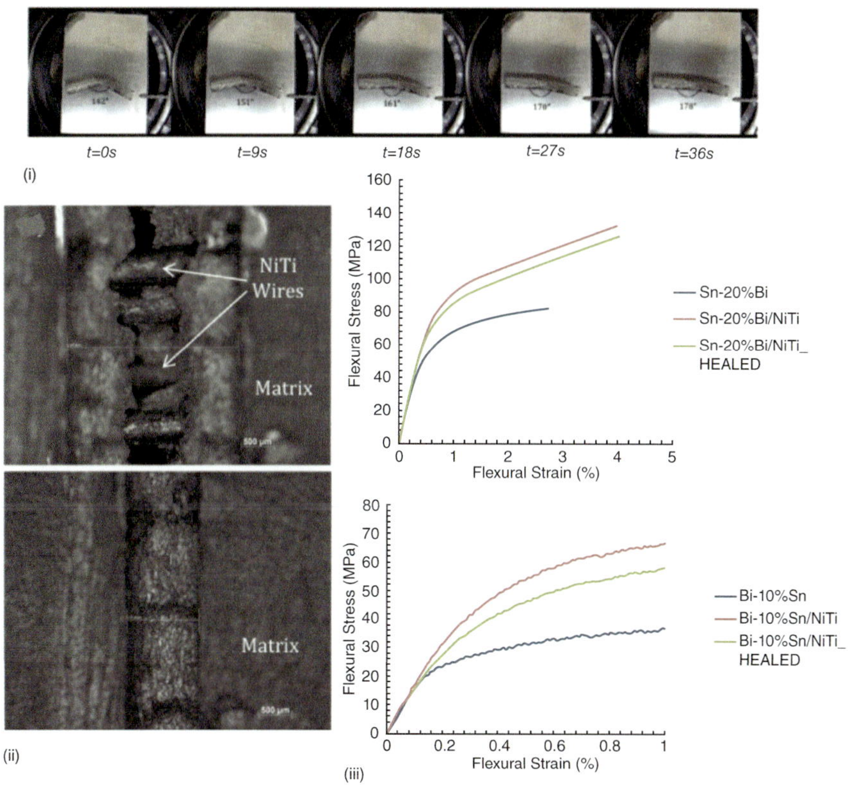

FIGURE 7.2 (i) Sequential recovery of the bent Sn–20 wt.% Bi–NiTi composite during the healing process, illustrating the shape restoration over time.[10] (ii) Crack interface showing visible narrowing due to the shape memory effect of embedded NiTi fibers.[10] (iii) Flexural strength curves for Sn–20 wt.% Bi–NiTi and Bi–10 wt.% Bi–NiTi self-healing composites, highlighting the strength recovery after thermal healing.[10]

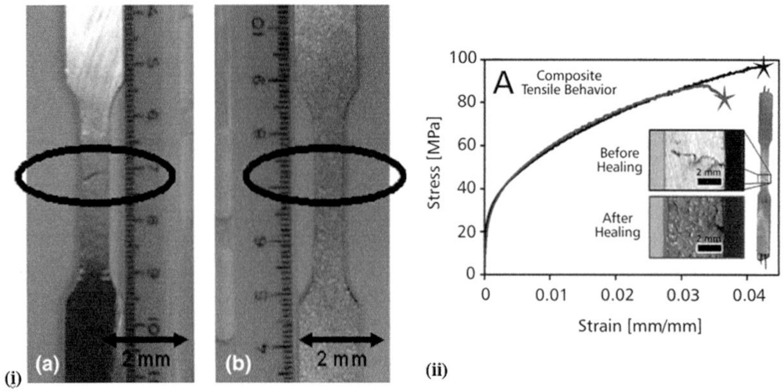

FIGURE 7.3 (i) (a) and (b) Sn–13 at.% Bi–NiTi composite subjected to tensile testing, showing complete crack closure after healing at 169°C. (ii) Tensile stress–strain plot illustrating the mechanical behavior of the composite and the recovery in tensile strength achieved post-healing.[39]

Within 1 hour, the Sn-20% Bi-NiTi composite recovered 92% of its original flexural strength, while the Bi-10% Sn-NiTi composite regained 88% (see Figure 7.2(iii)).

Wright et al. investigated fatigue crack healing in Al–Si, Al–Cu, and Al–Cu–Si alloys used as matrix materials, each reinforced with 2 vol.% shape memory alloy (SMA) wires.[36] The matrix was selected based on the design of even dispersion of low eutectic melting phase in the microstructure to have liquid-assisted healing behavior. Al-Si alloys show significant crack healing compared to Al-Cu and Al-Cu-Si alloy matrix retaining 90 % of ultimate tensile strength (UTS) after healing. This is mainly due to its well-connected eutectic phases in the microstructure and better ductility compared to the other alloys. Fisher[7,8] works on exploring different binary alloys (Al-Sn, Al-Cu, Al-Mg, Al-Si) and ternary alloys (Al-Cu-Si and Al-Mg-Li) to design and see how healing is affected by different alloying elements in aluminum. The work is based on Johnson Cooke constitutive model[40], a theoretical model developed by Manuel and further utilized by Fisher to estimate the volume % of SMA fiber required to provide sufficient clamping force to narrow or close the crack.[7,38] The Johnson-cook model was further simplified by solely considering the thermal effect and neglecting other factors such as strain hardening. The final expression includes controlling parameters such as healing temperature, thermal softening constant of the selected alloy as a matrix, compressive yield strength of matrix material and recovery stress of the SMA fiber. The proposed model has several limitations, particularly in its inability to accurately predict the minimum volume percentage of SMA fibers required to completely close or narrow the cracks in self-healing composites with higher temperature matrices particularly aluminum.

Ferguson[41] studied zinc-based matrix alloys and the effectiveness of both direct and indirect load transfer methods. In particular, mechanical load transfer was enhanced using embedded bolts in samples fabricated through permanent mold die casting. Rohatgi[25,26,42] introduced the concept of using SMA particles and fibers at macro- to nanoscales to achieve localized self-healing by filling nanovoids and nanocracks by generating compressive stresses during phase transformation of SMA particles. He also suggested various external triggers—such as electromagnetic, microwave, ultrasonic, and laser energy—and fabrication techniques like stir mixing, squeeze casting, and powder metallurgy for synthesizing SHMMCs.[25,26]

Poormir[43,44] investigated the influence of factors (i.e., pre-strain, healing temperature, vol. % of SMA) designed through Taguchi approach to check the healing in Sn-13 wt.% Bi alloy reinforced with Ni-Ti in a different form (continuous fiber and strip) under bending and tensile loading conditions. Zhu et al.[45,46] presented a simulation-based approach for three-dimensional metal matrix composites (MMCs) reinforced with shape SMA wires. The SMA behavior, including elasticity, phase transformation, and plasticity, is captured through a constitutive model. A method is introduced to incorporate pre-strained SMA wires within the composite to enhance crack closure. Finite element simulations, supported by experimental validation under three-point bending, reveal that SMA reinforcements significantly improve self-healing performance as shown in Figure 7.4. Both matrix softening and SMA pre-strain contribute to more effective crack closure. The developed model offers a reliable framework for designing MMCs with advanced self-healing capabilities for structural applications.

Till now the empirical study on selecting the suitable material and hit and trial approach to identify the parameters influencing self-healing in metal has limited the proper design of composites microstructure with its predictable self-healing behavior.

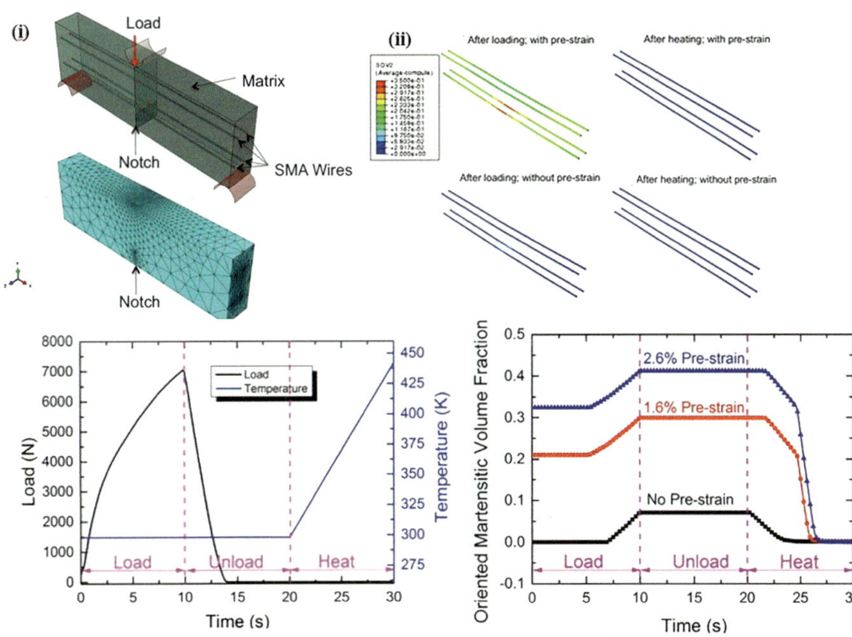

FIGURE 7.4 (i) Schematic of a three-point bending test simulation for a matrix embedded with SMA fiber, showing the meshing, applied loading and heating conditions. (ii) Effect of orientation of the martensitic phase in the fiber before and after heating under pre-strain conditions and without pre-strain.[45, 46]

At the macroscale, the mechanics and thermodynamics involved in SMA during phase transformation contribute to recovery stress for crack closure, while at the molecular level, understanding the flow of healing agents at the atomic scale is essential. Molecular dynamics simulations could offer valuable insights for designing effective self-healing composites by evaluating the field strain developed by distributed reinforced SMA fiber (particulates, nano fibers, continuous fibers) on matrix to narrow the crack or fill the microvoids present in the structure. Additionally, there is a need to develop a theoretical model capable of predicting the required volume fraction of SMA fibers that should include the effects of fiber diameter, pre-strain, arrangement, and matrix properties at the healing temperature to achieve complete crack closure or shape recovery. The efficacy mechanics-based theoretical model could be validated through advanced real-time monitoring systems during healing involving high resolution cameras, Digital image correlations, ultrasonic-based,[47–49] and impulse excitation-based advanced structural health monitoring techniques. Further study of flowing of liquid metals (as a part of matrix) to the crack could be tracked using advanced characterization, i.e., in situ SEM and 3-D X-Ray Tomography. This systematic study could help in designing a proper microstructure and resolve many unsolved mysteries of self-healing for single and multi-cycle healing in metals and metal-based composites.

7.2.2 Electro-pulse-Induced Healing in Metallic Systems

Electro-pulsing is the process associated with inducing high pulse electrical currents with high current density for a particular time interval. The high temperature heating and melting of crack interface to fill the crack is one of the conventional techniques that is being used to mend the crack. During conventional approach the identification of damaged location could not be spotted and therefore the drawback relies on that during high temperature the microstructure of undamaged area properties are degraded causing large Heat affected Zones.[50–52] Other targeted heating approaches like laser heating are very expensive and limited to surface cracks. Electro-pulsing techniques operate on the principle of electron transfer through a disordered lattice structure, which is disrupted by defects such as voids, microcracks, lattice distortions, and vacancies. The passage of high-density electrical pulses promotes dislocation movement and rearrangement, facilitating the healing of micro-damage. Additionally, localized Joule heating around macro-defects contributes to the collapse of microcracks and promotes micro-welding, thereby enhancing structural integrity[53,54] as shown in Figure 7.5(i). Electro-pulsing possesses an intrinsic capability to autonomously target damaged regions within a material. As pulsed current flows through the material, these high-resistance zones induce a detour effect, redirecting current density and concentrating Joule heating precisely at the damaged sites. This localized thermal and electrical stimulation facilitates selective healing, without affecting the undamaged regions.

Tao et al.[53] conducted works on SUS304 stainless steel to study the morphology of healed areas through SEM and EBSD analysis. The crack length of 6120 μm with a maximum width of 23 μm, decreasing partially through center on both the sides, was healed to a partial length to 1347.08 μm and 1287.42 μm from the middle to either side after seven times pulse discharge as shown in Figure 7.6. Crack healing was mainly due to detour and joule heating and shows namely three distinct healing areas,

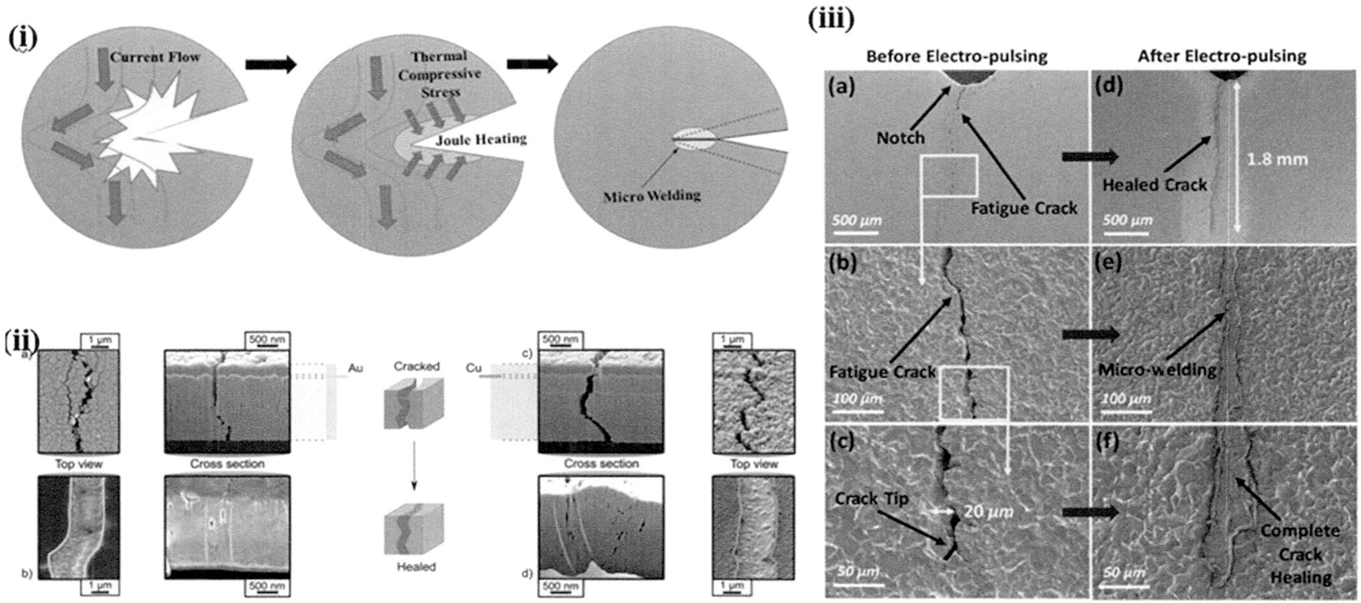

FIGURE 7.5 (i) Schematic illustration of the electro-pulse effect, highlighting the sequential steps contributing to micro-welding.[55] (ii) Top-view and cross-sectional images of Au and Cu thin films before and after healing, demonstrating crack closure via electro-pulse treatment.[56] (iii) SEM micrographs of steel showing a fatigue crack before and the healed region after electro-pulse treatment.[55]

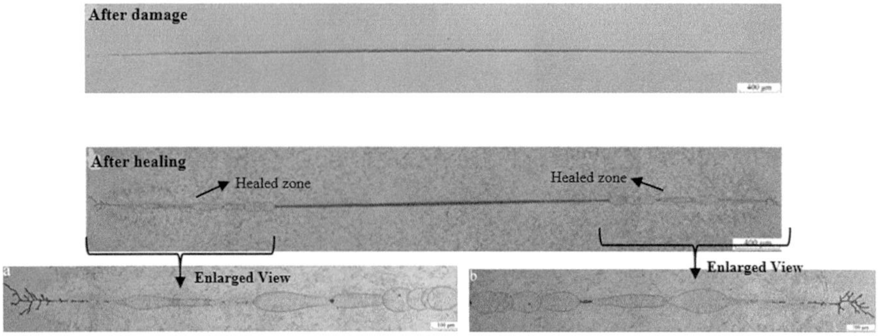

FIGURE 7.6 Morphology of the crack in SUS304 stainless steel before and after electro-pulse treatment, showing significant crack closure and surface modification.[53]

i.e., area A (columnar Crystal zone) caused due to high temperature, area B (recrystallization zone with fine grains) and area C (unaffected zone) due to low temperature generation. A DC electro-pulsing treatment carried out on aluminum (cold rolled 4043 and 2024 T4 heat treated) alloys with damage generated through tensile and fatigue were studied.[57] The change in resistance, yield, and ultimate tensile strength was evaluated for pulse frequency of 50 Hz and current density of 170 A/mm^2 for different pulsing treatment time durations and compared with as received, damage and electro-pulse treated samples. It is observed that the resistance (measured through electrical bridge) of the damaged samples increases and ultimately decreases after electro-pulse treatment. Moreover, the fatigue life of the electro-pulse treated sample at optimum time duration of 0.8 sec shows around 40 % of increase in fatigue life compared to untreated samples. In another study, the effect of electric pulse on fatigue crack healing in steel was analyzed and the microstructure near and away from the crack through SEM and EBSD analysis.[55] The pulse current density of 100 A/mm^2 for 10 sec pulse duration was applied to the damaged sample using pulse electric current generator. The pre-crack length of 1.8 mm and width of 20 μm is healed and micro-welding was observed between the cracked surface due to joule healing as shown in Figure 7.5(iii). Other reasons, i.e., the detouring effect, thermal compressive stress, and collapse of cracks, could be the potential reason for the fatigue crack healing.

Crack healing was studied in Ni/Al multilayered thin film 50 nm bilayer thickness with a metal layer of 100 nm thick film of either Au or Cu.[56] The crack opening ranging from 100 nm to 1.5 μm is locally activated through heat source obtained from a tungsten probe through DC power supply using a watch battery. The atomic diffusion of Ni atom into Al atoms forms an intermetallic forming reaction, which produces a stable increase in temperature between 1450°C and 1700°C sufficient enough to melt the metal layer to fill the crack. A maximum crack opening of 500 nm was healed in 10^{-3} milliseconds

7.2.3 Enabling Self-Healing through Electro-deposition

One of the main challenges in healing metals is the low diffusion rate of metal atoms in the solid phase at room temperature. As a result, effective diffusion requires high temperatures and significant energy input. The mobility of metal ions could be enhanced in electrolyte of aqueous solution or melted salt.[58] Electrochemistry involves reducing metal ions to the cathode (crack surface) to a metal atom through reduction reaction, helping it depositing to the crack at atomic level to form strong chemical bond with the cracked surface to repair it. Effective mass transportation and diffusion provide a low temperature (40°C–50°C) healing in metal.[54,59] Zheng[60] investigated crack healing in nickel plates using the electrodeposition process. Artificial cracks, 1.5 mm long with openings of 20–50 μm, were introduced at the center of nickel plates ranging in thickness from 100 to 300 μm. Both high and low throwing power electrolyte solutions were employed under varying current densities to evaluate their influence on healing effectiveness and required duration. Results showed that healing efficiency and strength recovery declined with increasing plate thickness; notably, the 300 μm thick samples exhibited only 33% healing efficiency, whereas the 100 μm thick samples achieved up to 96% efficiency. Similarly inspired by the bone, transport-mediated healing in nickel cellular metals coated with polymer (Parylene D to insulate the undamaged site) was used to demonstrate the healing on a strut structure.[61] Successful continuous nickel deposition was observed on the damaged section resulting in strength recovery from 23% in the unhealed specimens to 104% in samples healed with an energy input of 3500 J as shown in Figure 7.7.

7.2.4 Healing Mechanisms in Metals via Low-Melting-Point Alloys

In metals, the self-healing concept has been adapted from polymer-based self-healing materials, where healing agents such as DCPD, epoxy, or various monomers are encapsulated within microcapsules, microballoons, or hollow tubes vascular structure embedded in the polymer matrix.[16,17,20,62,63] When damage occurs, these capsules rupture, releasing the healing agent into the crack. This initiates a polymerization reaction, bonding the crack surfaces and restoring mechanical integrity.

For metals, one approach to achieving healing at relatively low temperatures involves the use of low-melting-point alloys—such as Sn, In, Zn, Sn-Bi, or Sn-Pb—which can flow into cracks upon heating and solidifying, thereby filling the damage through capillary force and surface tension and recovering structural strength[42] (see Figure 7.8).

Lucci et al.[34] used a 3D Ansys Fluent model to simulate the flow of low-melting alloys for thermal self-healing in metals. They introduced a healing factor, representing the ratio of viscous and surface forces to inertial forces, considering parameters like viscosity, contact angle, surface tension, density, flow velocity, and thermal expansion. Their results showed that when the healing factor is high, the healing agent stays within the microballoons and does not flow into the crack, limiting the healing effectiveness. Rohatgi et al.[25] conducted computational fluid dynamics (CFD) simulations to identify crack orientations and widths suitable for healing via liquid flow from microballoons. Their predictions, partially validated by experiments, showed that in Al-Zn SHMMCs, a 1 mm crack at 180° remained unfilled. In Al-Sn systems, reducing the crack width from 1 mm to 0.10 mm enabled full healing against gravity. For Sn-In SHMMCs, complete healing was achieved at 0° and 90° orientations when the crack diameter was 20% of the container, but not at 180° (see Figure 7.9).

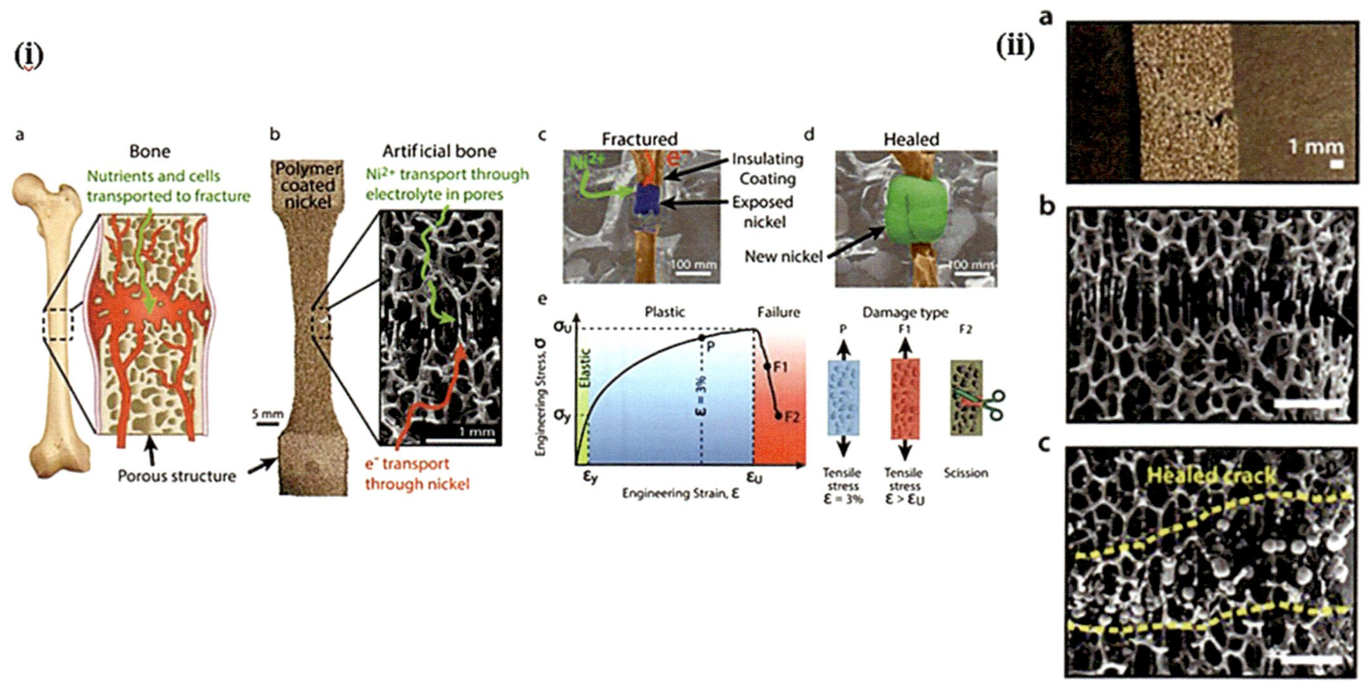

FIGURE 7.7 (i) (a) and (b) Bone-inspired cellular nickel structure illustrating the transport of nickel ions into the pores and their reduction to form nickel atoms. (c) and (d) Healing of a plastically deformed structure (at 3% strain) via electrodeposited nickel. (e) and (f) Deformation behavior of the cellular nickel structure under different loading modes.[61] (ii) (a) Cellular nickel structure damaged under tensile loading near failure point F1. (b) and (c) SEM micrographs showing the damaged region and the healed section after electrodeposition.

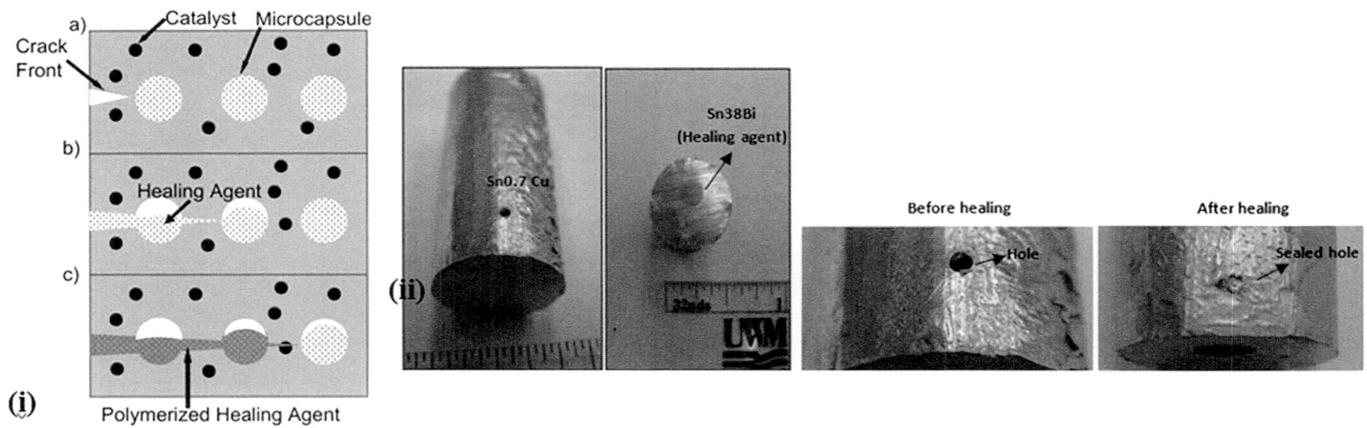

FIGURE 7.8 (i) Schematic illustration showing crack healing through the rupture of microcapsules containing a healing agent, which releases and fills the crack. (ii) Cross-sectional view of a Sn–0.7Cu alloy matrix embedded with Sn–38Bi healing agent, demonstrating effective sealing of a hole after the healing process.[42]

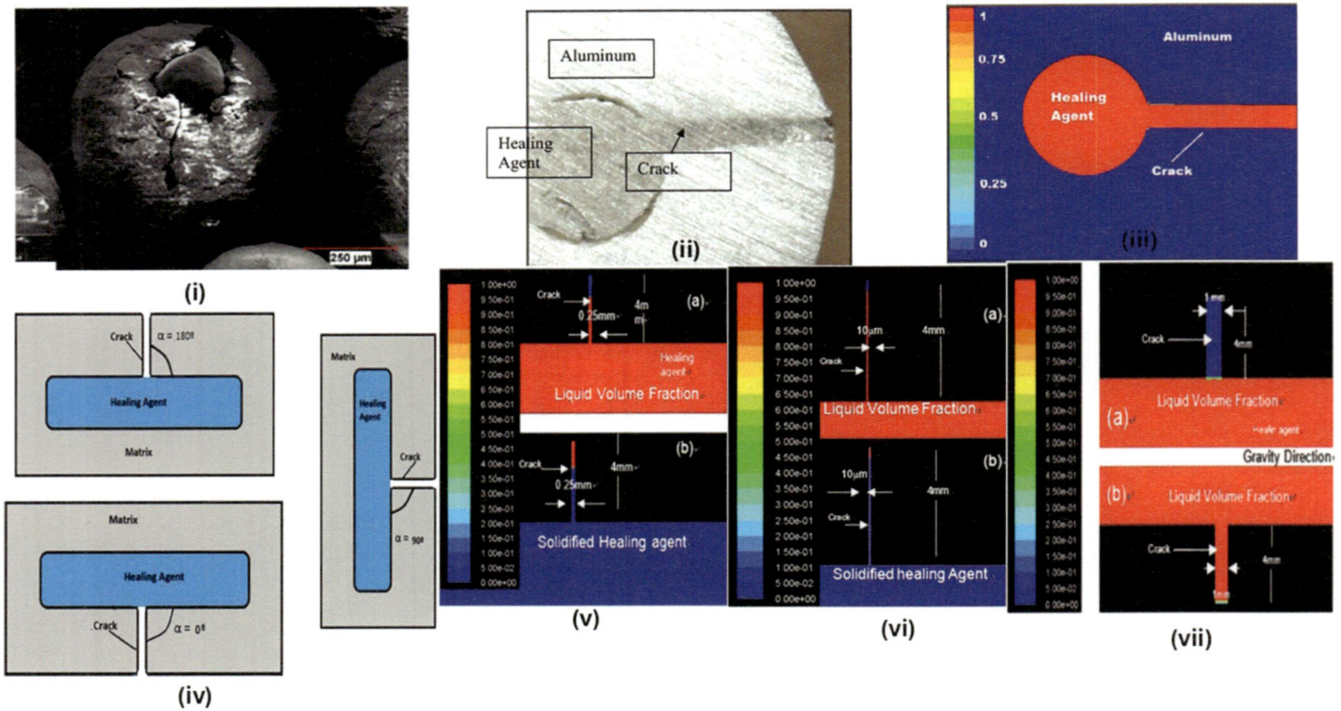

FIGURE 7.9 (i) SEM image of an alumina micro balloon filled with Sn–Bi eutectic alloy.[64] (ii) and (iii) Comparison of experimental and simulation results for the Al–Sn system with a 90°-oriented crack, showing crack filling up to 20% of the container diameter.[65] (iv) Schematic representation of crack orientation used in both simulation and experimental validation.[34, 65] (v) Healing agent filling in 0.25 mm and 4 mm wide cracks oriented at 90°. (vi) Crack filling behavior for 10 µm and 4 mm wide cracks oriented at 90°.[65] (vii) Crack filling in 1 mm diameter cracks oriented both along and against the direction of gravity.[34, 65]

Rohatgi et al.[26,29,30,42,66] at the University of Wisconsin-Milwaukee developed self-healing composites by incorporating porous alumina microballoons filled with 42Sn-58Bi solder into a 99.3Sn-0.7Cu matrix using a pressure infiltration process. In 2015, Alaneme et al.[67] experimentally demonstrated the self-healing capability of Al-Mg-Si alloy by embedding a central column of low-melting 60Sn-40Pb alloy. After pre-cracking and thermal activation, a healing efficiency of 91% was achieved, attributed to the continuous contact and potential bonding between the healing alloy and the matrix. Srivastava et al.[68] developed hybrid SHMMCs by embedding 0.48 mm NiTi wires (1 vol.%) and 6 vol.% Sn60Pb40 alloy introduced through a column longitudinally in AA2014 matrix samples. After inducing partial and full flexural damage, samples were heat-treated at various temperatures. Optimal healing occurred at 600°C, restoring 80% shape recovery and 52% flexural strength in partially damaged samples. In a follow-up study,[69] the Taguchi L8 orthogonal array was used to evaluate how NiTi wire vol. %, wire diameter, and specimen size affect healing in hybrid SHMMCs. Using an AA2014 matrix with Sn60Pb40 solder and varying SMA configurations, healing duration was identified as the most critical factor. The study achieved nearly 100% crack width closure, 96.95% depth recovery, and 73.76% flexural strength restoration. There have also been instances where microstructures were designed in such a way that the higher melting point alloy matrix materials contain low-melting-point alloys with varying concentration of Sn and Bi as (Al81Cu13Si6)100$-x$(Sn57Bi43)x where $x = 0$, 1 and 3 wt. % to form a composite.[70] During warm rolling at 149.85°C the Sn-Bi rich element, which has high mobility, diffuses toward the high-strain region filling the crack gap more effectively.

7.3 CURRENT CHALLENGES AND RESEARCH GAPS

To successfully achieve self-healing functionality through shape memory alloy-assisted self-healing in metals and metal based composites, several critical design factors and constraints must be considered: adopting a processing technique that ensures defect-free samples; selecting an alloy with a microstructure rich in eutectic composition; ensuring good wetting between the SMA fibers and the matrix to facilitate effective load transfer; optimizing the volume fraction of SMA reinforcement; maximizing the recovery stress of the SMA fibers; selecting the appropriate fiber diameter; strategically arranging the fibers within the matrix; and choosing a healing temperature that aligns with the properties of the matrix alloy. The core purpose of developing self-healing materials is to extend their usability and reduce the need for replacement, thereby conserving energy and resources. However, if healing requires high temperatures, it undermines this purpose—both by consuming significant energy and by potentially degrading the material itself. The energy saved by using self-healing materials is wasted if too much is spent on heating, making the process less practical and environmentally sustainable. Electro-pulsing treatment in metals, while promising for selectively healing microcracks, presents several limitations. From the past results it is observed partial crack closure is obtained, particularly for longer or wider cracks, and induces microstructural heterogeneity due to localized heating, leading to distinct zones with varying grain structures and mechanical properties. The effectiveness of EPT is highly sensitive to process parameters such as current density, pulse duration, and frequency, necessitating precise optimization to avoid under- or over-treatment. Additionally, the technique is less effective for large or deep cracks, especially in thin films, and although it aims to target only damaged areas, excessive Joule heating can still inadvertently affect surrounding undamaged regions. The process is also highly material-dependent, requiring case-specific calibration, and the need for specialized equipment and careful control limits its scalability for industrial use. Despite claims of autonomous targeting through the detour effect, achieving full autonomy in complex or large-scale structures remains challenging. Importantly, while many studies report microstructural changes and crack closure, there is a notable very few data on actual recovery in mechanical strength (e.g., tensile or flexural strength) after EPT, raising concerns about the true restoration of load-bearing capacity. Furthermore, the setup cost for implementing electro-pulsing is high, as it requires a robust pulse generator system equipped with several capacitor banks for rapid charging and discharging to generate high-density pulses, making the technique expensive and complex to deploy at scale. The electrodeposition process for metal crack healing faces several limitations. Healing efficiency significantly decreases with increasing metal thickness, making it less effective for thicker components. The process requires precise control over electrolyte composition and current density to achieve optimal results, which can be challenging. Although healing occurs at relatively low temperatures, it often demands substantial energy input and extended deposition times. Additionally, electrodeposition is limited to cracks accessible by the electrolyte, restricting its use for internal or hard-to-reach defects. Uneven metal deposition may occur without careful regulation, potentially compromising the quality of the repair. Furthermore, to ensure selective healing, undamaged areas must be insulated, adding complexity to the treatment process. A significant limitation for aluminum is its rapid formation of a stable oxide layer (aluminum oxide) when exposed to aqueous environments, which passivates the surface and prevents effective electrodeposition in conventional aqueous electrolytes.[71–73] As a result, aluminum cannot be easily electrodeposited from standard aqueous solutions and typically requires specialized high-temperature ionic liquid electrolytes, restricting its practical application for electrodeposition-based healing.[74–76] Moreover, scaling up the electrodeposition process to large samples necessitates proportionally larger electrolyte containers and more complex setups, increasing costs and logistical challenges. While low-melting-point alloys present a promising approach for achieving self-healing in metallic systems, their integration into metal matrices comes with several critical limitations. One of the primary challenges is the complexity of embedding microballoons or microcapsules uniformly within the matrix. This often leads to a non-uniform distribution of healing agents, where cracks may develop in regions devoid of microcapsules or micro ballons—preventing rupture and thereby inhibiting the release of the healing alloy into the damaged site. As a result, healing becomes inconsistent and highly location dependent. Moreover, these systems are typically limited to single-cycle healing. Once the healing agent flows out and solidifies within the crack, it leaves behind voids at the original location, reducing structural integrity and making repeated healing cycles unfeasible without further processing. Another significant drawback is the reduction in mechanical strength of the composite. Low-melting-point alloys such as Sn-Bi or Sn-Pb possess low strength compared to the primary matrix, and their incorporation introduces material anisotropy, which can act as stress concentrators and reduce durability under cyclic or high-load conditions. These challenges highlight the need for further refinement before such self-healing methods can be reliably employed in structural or critical applications.

7.4 OUTLOOK AND FUTURE RESEARCH

The advancement of truly autonomous and repeatable self-healing mechanisms in metal matrix composites remains a pressing challenge in material science. Most existing approaches rely on external stimuli such as thermal input or electrical current and typically demonstrate only a single healing cycle. To overcome these limitations, it is essential to develop self-contained systems capable of autonomous, multi-cycle healing, particularly ones that can function at or near room temperature.

A promising strategy involves integrating thermite-based chemical systems with encapsulated healing agents. When a crack ruptures these containers, the thermite reacts exothermically with atmospheric oxygen, releasing intense heat. This localized heat generation is sufficient not only to melt low-melting-point alloys stored within micro- or nano-containers but also to partially liquefy the surrounding matrix, enabling effective sealing of the crack through re-solidification. This approach holds significant potential for creating self-welding mechanisms within metallic systems.

The use of nanoscale capsules or nanotubes filled with low-melting-point alloys offers additional advantages. These nanocontainers can respond to damage at the microscale or even nanoscale, sealing cracks before they grow larger. Moreover, the incorporation of such nanofillers can improve both the mechanical properties of the matrix and the healed material. However, the reliable filling of nanotubes with eutectic or low-melting-point alloys before their integration into the matrix remains a critical challenge that warrants further investigation. Negative thermal expansion (NTE) materials such as ZrW_2O_8, $HfMo_2O_8$, and $Sc_2(WO_4)_3$ can be explored as reinforcements in metal matrix composites to facilitate nanoscale crack closure during low to moderate heating.[77,78] These materials exhibit contraction upon heating, which can help pull crack faces together and enhance the effectiveness of self-healing, particularly at the early stages of damage. Although ZrW_2O_8 is known for its isotropic negative thermal expansion over a wide range (0°C to 1050°C), it remains effective even at temperatures below 200°C, making it compatible with low-temperature healing strategies. For SHMMCs designed to operate at or below 200°C, incorporating such NTE materials can support crack narrowing without requiring high thermal input. This synergy is especially beneficial when combined with low-melting healing agents or SMAs activated at moderate temperatures.

The shape memory effect of embedded SMA fibers offers an efficient mechanism for crack narrowing. When activated, these fibers exert recovery stress that can bring crack faces into contact, promoting closure. A detailed and accurate theoretical model is needed to simulate the resulting strain fields, accounting for both the SME and the matrix's deformation characteristics. Integrating the partially developed theoretical model with Physics-based machine learning algorithm will help in refining the theoretical model further for an accurate prediction of self-healing properties after healing.

Another essential component of SHMMC development is the incorporation of structural health monitoring (SHM). This enables real-time tracking of damage propagation and healing. Techniques such as strain imaging, X-ray tomography, CT scanning, and non-destructive evaluation using piezoelectric sensors, ultrasonic probes, and electromechanical impedance measurements can be employed. Furthermore, the effect of randomly oriented SMA fibers on healing cracks of various shapes, sizes, and orientations should be thoroughly assessed through both experiments and simulations.

The flow behavior of healing agents from ruptured capsules, nanotubes, or eutectic reservoirs into cracks must also be studied in-depth. Using CFD, researchers can quantify the effects of surface tension, capillarity, gravity, and crack geometry on the infiltration and subsequent solidification of healing agents. This will help predict the healing efficiency under various damage scenarios. The impact of different types of healing agents—ranging from low-melting alloys and solders to polymers—on both crack propagation resistance and post-healing mechanical performance also requires systematic exploration.

One of the major challenges in metal-based self-healing is avoiding oxidation during high-temperature healing, as oxides can inhibit proper crack closure. The viability of adding suitable alloying elements—such as chromium, aluminum, nickel, and cerium—that could reduce oxide formation on the surfaces of cracks should be investigated as a future work. Additionally, heating the entire component is impractical, particularly for large-scale structures. Therefore, extensive research is required to advance both electro-pulsing and electrodeposition techniques for self-healing applications. In particular, developing a reliable electrodeposition process for direct crack filling using the parent metal, especially aluminum substrates, remains a critical area of focus.

Finally, developing additive manufacturing (AM) technologies, such as 3D printing, to offer potential routes to fabricate multi-material systems containing embedded healing agents, capsules, or SMA reinforcements could be a major breakthrough. AM could provide precise spatial control over the placement of these functional elements with controlled parameters to fabricate self-healing metallic composite materials. Continued interdisciplinary research and collaborative innovation will be key to overcoming current limitations and enabling broader adoption across industries.

REFERENCES

1. Calvin, K. et al. *IPCC, 2023: Climate change 2023: Synthesis report*. Contribution of Working Groups I, II and III to the Sixth Assessment Report of the Intergovernmental Panel on Climate Change [Core Writing Team, Lee, H. & Romero, J. (eds.)]. Geneva: IPCC. https://www.ipcc.ch/report/ar6/syr/ (2023) doi:10.59327/IPCC/AR6-9789291691647
2. Department of Economic and Social Affairs. Self healing, enzymatic, bio-concrete to fight climate change. https://sdgs.un.org/partnerships/self-healing-enzymatic-bio-concrete-fight-climate-change
3. Geyer, R., Jambeck, J. R. & Law, K. L. Production, use, and fate of all plastics ever made. *Sci. Adv.* **3**, e1700782 (2017).

4. Cheng, Y., Wang, C., Kang, F. & He, Y.-B. Self-healable lithium-ion batteries: A review. *Nanomaterials* **12**, 3656 (2022).
5. Self-Healing Materials Market Size & Share Report, 2030, 102. https://www.grandviewresearch.com/industry-analysis/self-healing-materials
6. Shen, R., Amano, R. S., Lewinski, G. & Matt, A. K. K. A new vascular system highly efficient in the storage and transport of healing agent for self-healing wind turbine blades. *J. Energy Resour. Technol.* **141** (2019).
7. Fisher, C. R. *Design of liquid-assisted self-healing metal-matrix composites.* Florida: University of Florida, 2013.
8. Fisher, C. R. et al. Repairing large cracks and reversing fatigue damage in structural metals. *Appl. Mater. Today* **13**, 64–68 (2018).
9. Rohatgi, P. Self healing lead, tin, and their alloys, and solders, incorporating shape memory alloys, reactive particles and hollow vascular networks. U.S. Patent No. 10,161,026. 25 Dec. 2018.
10. Mishra, S. K. *Shape memory alloy reinforced self-healing metal matrix composites.* Milwaukee: University of Wisconsin-Milwaukee, 2013.
11. Kim, S., Jeon, H., Koo, J. M., Oh, D. X. & Park, J. Practical applications of self-healing polymers beyond mechanical and electrical recovery. *Adv. Sci.* **11**, 2302463 (2024).
12. Ghosh, S. K. *Self-healing materials – Fundamentals, design strategies and applications.* Weinheim: Wiley-VCH Verlag GmbH, 2008.
13. Benefits and Applications of Self-Healing Materials. https://www.techsciresearch.com/blog/benefits-and-applications-of-self-healing-materials/4511.html (2024).
14. El Choufi, N., Mustapha, S., Tehrani B., A. & Grady, B. P. An overview of self-healable polymers and recent advances in the field. *Macromol. Rapid Commun.* **43**, 2200164 (2022).
15. Barbero, E. J., Ford, K. J. & Mayugo, J. A. Modeling self-healing of fiber-reinforced polymer-matrix composites with distributed damage. In *Self-healing materials*, 267–283. John Wiley & Sons, Ltd, 2008. doi:10.1002/9783527625376.ch9
16. Zhang, M. Q., Rong, M. Z. & Yin, T. Self-healing polymers and polymer composites. In *Self-healing materials*, 29–71. John Wiley & Sons, Ltd, 2008. doi:10.1002/9783527625376.ch2
17. Cerdan, K. et al. Sustainability of self-healing polymers: A holistic perspective towards circularity in polymer networks. *Prog. Polym. Sci.* **152**, 101816 (2024).
18. Chowdhury, S., Ray, R. & Ghosh, P. S. Self-healing concrete – A potential sustainable solution for durable concrete structures. Tech Speak, Tata Consulting Engineers' Blog (2023). https://www.tce.co.in/blogs/self-healing-concrete-a-potential-sustainable-solution-for-durable-concrete-structures/
19. Nakao, W., Takahashi, K. & Ando, K. Self-healing of surface cracks in structural ceramics. In *Self-healing materials*, 183–217. John Wiley & Sons, Ltd, 2008. doi:10.1002/9783527625376.ch6
20. Blaiszik, B. J. et al. Self-healing polymers and composites. *Annu. Rev. Mater. Res.* **40**, 179–211 (2010).
21. Bellah, M., Nosonovsky, M. & Rohatgi, P. Recent advances in self-healing metal matrix composites. In *Metal-matrix composites* (eds. Srivatsan, T. S., Rohatgi, P. K. & Hunyadi Murph, S.), 297–310. Cham: Springer International Publishing, 2022. doi:10.1007/978-3-030-92567-3_18
22. Rohatgi, P. K. Al-shape memory alloy self-healing metal matrix composite. *Mater. Sci. Eng. A* **619**, 73–76 (2014).
23. van Dijk, N. & van der Zwaag, S. Self-healing phenomena in metals. *Adv. Mater. Interfaces* **5**, 1800226 (2018).

24. Trask, R. S., Williams, H. R. & Bond, I. P. Self-healing polymer composites: Mimicking nature to enhance performance. *Bioinspir. Biomim.* **2**, P1–P9 (2007).
25. Rohatgi, P. Self-healing aluminum alloys incorporating shape metal alloys and reactive particles. U.S. Patent No. 9,435,014. 6 Sep. 2016.
26. Rohatgi, P. K. Self healing metals and alloys—including structural alloys and self-healing solders. U.S. Patent No. 8,518,531. 27 Aug. 2013.
27. Srivastava, V. & and Gupta, M. Design and prediction of healing assessment for AA2014-Nitinol strip-Solder alloy based hybrid self-healing composite structure via Taguchi analysis and fuzzy logic approach. *Mech. Adv. Mater. Struct.* **30**, 4196–4215 (2023).
28. Bellah, M., Nosonovsky, M. & Rohatgi, P. Shape memory alloy reinforced self-healing metal matrix composites. *Appl. Sci.* **13**, 6884 (2023).
29. Nosonovsky, M. & Rohatgi, P. K. Thermodynamic principles of self-healing metallic materials. In *Biomimetics in materials science: Self-healing, self-lubricating, and self-cleaning materials* (eds. Nosonovsky, M. & Rohatgi, P. K.), 25–51. New York: Springer, 2012. doi:10.1007/978-1-4614-0926-7_2
30. Nosonovsky, M. & Rohatgi, P. K. Case study of self-healing in metallic composite with embedded low melting temperature solders. In *Biomimetics in materials science: Self-healing, self-lubricating, and self-cleaning materials* (eds. Nosonovsky, M. & Rohatgi, P. K.), 53–73. New York: Springer, 2012. doi:10.1007/978-1-4614-0926-7_3
31. Kumar, P. K. & Lagoudas, D. C. Introduction to shape memory alloys. In *Shape memory alloys: Modeling and engineering applications*, 1–51. Boston, MA: Springer US, 2008. doi:10.1007/978-0-387-47685-8_1
32. Srivastava, V. & Gupta, M. Impact of post hardening mechanism on self-healing assessment of AA2014 nitinol-based smart composites. *Met. Mater. Int.* **27**, 2666–2681 (2021).
33. Salowitz, N., Misra, S., Haider, M. I., Povolo, M. & Rohatgi, P. Investigation into the performance of NiTi shape memory alloy wire reinforced Sn-Bi self-healing metal matrix composite. *Materials* **15**, 2970 (2022).
34. Martínez Lucci, J., Amano, R. S. & Rohatgi, P. K. Heat transfer and fluid flow analysis of self-healing in metallic materials. *Heat Mass Transf.* **53**, 825–848 (2017).
35. Manuel, M. V. Principles of self-healing in metals and alloys: An introduction. In *Self-healing materials*, 251–265. John Wiley & Sons, Ltd, 2008. doi:10.1002/9783527625376.ch8
36. Wright, M. C., Manuel, M. & Wallace, T. Fatigue resistance of liquid-assisted self-repairing aluminum alloys reinforced with shape memory alloys. https://ntrs.nasa.gov/citations/20140013299 (2013).
37. Manuel, M. V., Jr, Fisher, C. R. & Wright, M. C. Self-repairing metal alloy matrix composites, methods of manufacture and use thereof and articles comprising the same. U.S. Patent No. 10,597,761. 24 Mar. 2020.
38. Manuel, M. V. *Design of a biomimetic self-healing alloy composite* (Doctoral dissertation). Northwestern University, Washington, DC.
39. Manuel, M. V. & Olson, G. B. Biomimetic self-healing metals. *Proceedings of the 1st International Conference on Self-Healing Materials* (2007).
40. He, B. et al. Comprehensive shape memory alloys constitutive models for engineering application. *Mater. Des.* **225**, 111563 (2023).
41. Ferguson, J. B., Schultz, B. F. & Rohatgi, P. K. Zinc alloy ZA-8/shape memory alloy self-healing metal matrix composite. *Mater. Sci. Eng. A* **620**, 85–88 (2015).
42. Martinez Lucci, J., Amano, R. S., Rohatgi, P. & Schultz, B. Self-healing in an aluminum alloy reinforces with microtubes. *Proceedings of the Energy Nanotechnology International Conference*, 79–88. New York: American Society of Mechanical Engineers. American Society of Mechanical Engineers Digital Collection, 2009. doi:10.1115/ENIC2008-53011

43. Poormir, M. A., Khalili, S. M. R. & Eslami-Farsani, R. Investigation of the self-healing behavior of Sn-Bi metal matrix composite reinforced with NiTi shape memory alloy strips under flexural loading. *JOM* **70**, 806–810 (2018).
44. Poormir, M. A. & Farsani, S. M. R. Optimal design of a bio-inspired self-healing metal matrix composite reinforced with NiTi shape memory alloy strips. *Compos Struct.* **29**, 3972–3982 (2018).
45. Pingping, Z. *Shape memory alloy modeling and applications to porous and composite structures*. Washington, DC: Northwestern University, 2015.
46. Zhu, P. et al. Characterization and modeling of three-dimensional self-healing shape memory alloy-reinforced metal-matrix composites. *Mech. Mater.* **103**, 1–10 (2016).
47. Brandon, E. J. et al. Structural health management technologies for inflatable/deployable structures: Integrating sensing and self-healing. *Acta Astronaut.* **68**, 883–903 (2011).
48. Haider, M. I., Srivastava, V., Rohatgi, P. K. & Salowitz, N. P. Design of self-healing materials for the integration of structural health monitoring (SHM^2). *E-J. Nondestruct. Test.* **29**, 1–10 (2024).
49. Salowitz, N. P., Rohatgi, P. K., Suri, S. & Correa, A. L. Initial integration of structural health monitoring of self-healing materials. *E-J. Nondestruct. Test.* **21**, 1–8 (2016).
50. Cai, Q. et al. Crack healing via electropulsing treatment applied to additive-manufactured TiC/316L stainless steel composites. *Mater. Lett.* **365**, 136410 (2024).
51. Zhang, H. et al. Evolution of grain boundary character distributions in a cold-deformed Nickel-based superalloy during electropulsing treatment. *J. Mater. Res. Technol.* **9**, 5723–5734 (2020).
52. Zhang, S. et al. Rapid crack healing and fatigue life extension in single-edge notched pure titanium via electropulsing treatment. *Int. J. Fatigue* **198**, 108967 (2025).
53. Yu, T., Deng, D., Wang, G. & Zhang, H. Crack healing in SUS304 stainless steel by electropulsing treatment. *J. Clean. Prod.* **113**, 989–994 (2016).
54. Song, H., Wang, Z., He, X. & Duan, J. Self-healing of damage inside metals triggered by electropulsing stimuli. *Sci. Rep.* **7**, 7097 (2017).
55. Kumar, A. & Paul, S. K. Healing of fatigue crack in steel with the application of pulsed electric current. *Materialia* **14**, 100906 (2020).
56. Danzi, S. et al. Rapid on-chip healing of metal thin films. *Adv. Mater. Technol.* **4**, 1800468 (2019).
57. Shengru, Q., Yanli, L., Yun, L. & Chengyu, Z. Damage healing of aluminum alloys by D. C. electropulsing and evaluation by resistance. *Rare Met. Mater. Eng.* **38**, 570–573 (2009).
58. Mahmoudi, M., Oh, S. & Jolandan, M. M. Low energy repair of co-continuous metal-ceramic composites using electrodeposition. *J. Compos. Mater.* **57**, 593–603 (2023).
59. Kamble, C., Jadhav, V. V., & Mane, R. S. (2023). Electrodeposition of metal oxide nanostructures. In *Solution methods for metal oxide nanostructures* (pp. 197–220). Elsevier.
60. Zheng, X. G., Shi, Y.-N. & Lu, K. Electro-healing cracks in nickel. *Mater. Sci. Eng. A* **561**, 52–59 (2013).
61. Hsain, Z. & Pikul, J. H. low-energy room-temperature healing of cellular metals. *Adv. Funct. Mater.* **29**, 1905631 (2019).
62. Cheng-Hui, L. & Jing-Lin, Z. Self-healing polymers based on coordination bonds. *Adv. Mater.* Wiley Online Libr. (2020). doi:10.1002/adma.201903762
63. Amano, R. S., Lewinski, G. & Shen, R. Imprinted glass fiber-reinforced polymer vascular networks for creating self-healing wind turbine blades. *J. Energy Resour. Technol.* **144**, 062107 (2021).
64. Ruzek, A. *Synthesis and characterization of metallic systems with potential for self-healing*. Milwaukee: University of Wisconsin Milwaukee, 2009.
65. Lucci, J. M. *Heat Tansfer and fluid flow analysis of self healing in metallic materials*. Milwaukee: University of Wisconsin Milwaukee, 2011.

66. Amano, R., Rohatgi, P., Lucci, J. M., Schultz, B. & Ruzek, A. Design and demonstration of self-healing behavior in a lead-free solder alloy. In *7th International Energy Conversion Engineering Conference (American Institute of Aeronautics and Astronautics)*. doi:10.2514/6.2009-4514
67. Alaneme, K. K. & Omosule, O. I. Experimental studies of self healing behaviour of under-aged Al-Mg-Si alloys and 60Sn-40Pb alloy reinforced aluminium metal-metal composites. *J. Miner. Mater. Charact. Eng.* **3**, 1–8 (2015).
68. Srivastava, V. & Gupta, M. Experimental assessment of self-healing characteristics in AA2014 matrix with nitinol wire and solder alloy as healing agents. *Mater Res Express* **6**(8), 085704 (2019). IOP Science.
69. Srivastava, V. & Gupta, M. Experimental assessment of self-healing nature in aluminum-based smart composites with NiTi wires and solder alloy as healing agents through Taguchi approach. *Compos Struct.* **31** (2020).
70. Kim, J. et al. Thermally-triggered dual in-situ self-healing metallic materials. *Sci. Rep.* **8**, 2120 (2018).
71. Bakkar, A. & Neubert, V. A new method for practical electrodeposition of aluminium from ionic liquids. *Electrochem. Commun.* **51**, 113–116 (2015).
72. Maniam, K. K. & Paul, S. A review on the electrodeposition of aluminum and aluminum alloys in ionic liquids. *Coatings* **11**, 80 (2021).
73. Jiang, T., Chollier Brym, M. J., Dubé, G., Lasia, A. & Brisard, G. M. Electrodeposition of aluminium from ionic liquids: Part I—electrodeposition and surface morphology of aluminium from aluminium chloride (AlCl3)–1-ethyl-3-methylimidazolium chloride ([EMIm]Cl) ionic liquids. *Surf. Coat. Technol.* **201**, 1–9 (2006).
74. Shi, M., Jiang, J. & Zhao, H. Electrodeposition of aluminum in the 1-Ethyl-3-methylimidazolium tetrachloroaluminate ionic liquid. *Electrochem* **2**, 185–196 (2021).
75. Wu, J. et al. Ionic liquids as electrolytes in aluminum electrolysis. *Front. Chem.* **10** (2022).
76. Guinea, E. et al. Robust aluminum electrodeposition from ionic liquid electrolytes containing light aromatic naphta as additive. *ChemistryOpen* **8**, 1094–1099 (2019).
77. Greve, B. K. et al. Pronounced negative thermal expansion from a simple structure: Cubic ScF3. *J. Am. Chem. Soc.* **132**, 15496–15498 (2010).
78. Attfield, M. P. Strong negative thermal expansion in siliceous faujasite. *Chem. Commun.* 601–602 (1998) doi:10.1039/A707141H

Index

Pages in *italics* refer to figures and pages in **bold** refer to tables.

For Product Safety Concerns and Information please contact our EU representative GPSR@taylorandfrancis.com Taylor & Francis Verlag GmbH, Kaufingerstraße 24, 80331 München, Germany